SAS Publishing

Lawrence H. Muhlbaier, Ph.D. - (S190)
Biometry & Medical Informatics
Box 3865
Duke University Medical Center
Durham, N. C. 27710

The Power to Know.

The correct bibliographic citation for this manual is as follows: Lafler, Kirk Paul. 2004. *PROC SQL: Beyond the Basics Using SAS®*. Cary, NC: SAS Institute Inc.

PROC SQL: Beyond the Basics Using SAS®

Copyright © 2004, SAS Institute Inc., Cary, NC, USA

ISBN 1-59047-534-8

All rights reserved. Produced in the United States of America. No part of this publication may be reproduced, stored in a retrieval system, or transmitted, in any form or by any means, electronic, mechanical, photocopying, or otherwise, without the prior written permission of the publisher, SAS Institute Inc.

U.S. Government Restricted Rights Notice: Use, duplication, or disclosure of this software and related documentation by the U.S. government is subject to the Agreement with SAS Institute and the restrictions set forth in FAR 52.227-19, Commercial Computer Software-Restricted Rights (June 1987).

SAS Institute Inc., SAS Campus Drive, Cary, North Carolina 27513.

1st printing, September 2004

SAS Publishing provides a complete selection of books and electronic products to help customers use SAS software to its fullest potential. For more information about our e-books, e-learning products, CDs, and hard-copy books, visit the SAS Publishing Web site at **support.sas.com/pubs** or call 1-800-727-3228.

SAS® and all other SAS Institute Inc. product or service names are registered trademarks or trademarks of SAS Institute Inc. in the USA and other countries. ® indicates USA registration.

Other brand and product names are trademarks of their respective companies.

Contents

Preface

Most books on this topic are written for users with little or no experience. These books generally address the basics quite well by presenting syntax and concepts with a sprinkling of examples here and there, but they often fall short of the needs of the power user. A basic understanding of SQL is not what all users want or need. Moreover, with the growing popularity and use of SQL, many users quickly master the basics and are left with nowhere to turn for learning the more advanced topics and features of the language.

I wrote this book to highlight the many features offered by PROC SQL, while providing information about the American National Standards Institute (ANSI) guidelines, as well as non-ANSI features. This book includes features defined in the ANSI standard, plus additional features that have been implemented in PROC SQL. It is intended to serve as a resource to complete users' mastery of the language while enabling them to perform more complicated tasks.

I also wanted to address the needs of the broad range of SAS users (from intermediate to advanced), including existing SQL users, application programmers, database designers and administrators, statisticians, and systems analysts. In taking this approach, I concentrated on the various ways PROC SQL is used to solve real application problems. This book is intended as a reference for SAS users who want to understand how PROC SQL can be used to perform common, and not so common, tasks. This book complements the *SAS SQL* documentation very nicely.

This is a book for intermediate and advanced SAS users who want clearly written explanations, examples, and techniques on the SQL procedure, all organized in a helpful and logical way. It is organized so users can get faster, easier, and better results with the SQL procedure without spending substantial amounts of time reading paragraph after paragraph of text or attempting to apply esoteric syntax-oriented examples. This book is the perfect companion for SAS and PROC SQL users who want step-by-step instructions and example code to enhance their data access, manipulation, and presentation prowess.

How This Book Is Organized

This book is organized so you can read it cover to cover in a linear fashion. Or, if a current project requires learning a particular subject, then you can use the index to find the help you need or simply skip around and read the chapters or sections of interest to you.

Starting with Chapter 1, "Designing Database Tables," readers are presented with database design strategies, database normalization principles, data redundancy techniques, and techniques for achieving optimal designs using first (1NF), second (2NF), and third (3NF) normal forms.

Chapter 2, "Working with Data in PROC SQL," presents a variety of operators, functions, and predicates that can be used while working with numeric and character data. Dictionary tables and the wonderful information they provide are also presented.

Chapter 3, "Formatting Output," presents techniques to improve output produced by queries. ANSI and non-ANSI techniques are presented to show how to insert text and constants between selected columns, how to display column headers for derived fields using mathematical expressions, how to display values in a certain order with the ORDER BY, GROUP BY, and HAVING clauses, and how the Output Delivery System (ODS) can be used to extend SQL's capabilities to work with output.

Chapter 4, "Coding PROC SQL Logic," examines how PROC SQL handles conditional logic operations including CASE and WHEN/THEN-ELSE. The availability of these programming constructs provides important features that typically are found only in host languages or external environments.

Chapter 5, "Creating, Populating, and Deleting Tables," explores topics such as creating tables using column definition lists and from other tables, implementing column and table integrity constraints as well as referential integrity, populating table rows, and deleting one or more tables.

Chapter 6, "Modifying and Updating Tables and Indexes," shows how to work with sets of data, modify database structures, rename tables, and define, use, and delete indexes.

Chapter 7, "Coding Complex Queries," presents the differences between merging and joining tables, techniques on combining two or more tables to form cross-referenced data, and using queries within queries. Joins, subqueries, and unions enable programmers to link one piece of information with another piece of information. Examples illustrate the different types of joins and set operators in PROC SQL.

Chapter 8, "Working with Views," examines the power of views and how "virtual" tables can eliminate or reduce data redundancy, shield users from making logic and/or data errors, hide sensitive information from users for security reasons, update data, and provide greater change control.

Chapter 9, "Troubleshooting and Debugging," presents a variety of debugging techniques including the use of various PROC SQL options, table options, and macro variables that can be used to troubleshoot SQL queries.

Chapter 10, "Tuning for Performance and Efficiency," presents techniques for improving a query's performance. A variety of techniques are presented, including the optimization of query expressions, avoiding unnecessary sorting, using care when applying indexes, creating effective joins, constructing efficient views, and eliminating second passes over data.

Reference Aids

This book provides the following features:

- Glossary: Contains descriptions of keywords and concepts.
- Comprehensive index: Includes references to useful task-oriented examples as well as references to PROC SQL keywords, terms, and concepts.
- Keyword and language references: PROC SQL keywords and the SAS language, referencing table and column names and excerpts of code. In most cases, this book exclusively uses uppercase for SAS code, with the exception of title information. SAS software permits code to be entered in lowercase, uppercase, or any mixture of the two. Titles and footnotes, if used, are displayed in output in the manner in which they were entered.

To my mother, father, and brother
for your love, support, and encouragement.

Acknowledgments

This book was made possible because of the support and encouragement of many people. I would like to extend my sincerest thanks to each person who encouraged me to write this book. As G. B. Stern once said, "Silent gratitude isn't very much use to anyone." So from me to you, thank you to each one of you.

To Art Carpenter of California Occidental Consultants for believing that a "beyond the basics" type of book on PROC SQL would be useful to SAS users. His support during the book's early stages is greatly appreciated. Art's support, encouragement, and suggestions during the development of the first draft helped bring *PROC SQL* to life.

To Julie Platt of SAS Institute for her guidance and editorial assistance during the editing process and for allowing me to see the light at the end of the long tunnel. Her encouragement and support gave me the inspiration to see this project to completion.

To Patsy Poole of SAS Institute for her support and encouragement during the development of the first draft and for coordinating the technical review process.

To Stephenie Joyner of SAS Institute for her support and encouragement during the development of the second, third, and final manuscript drafts, and for coordinating the technical review process.

To Paul Kent of SAS Institute for his contagious enthusiasm, great examples, clear explanations, and assistance with some of the fine points of PROC SQL and its capabilities over the years.

To David Baggett of SAS Institute for accepting the original *PROC SQL* manuscript over a decade ago (yes, it's been that long). Your encouragement and support over the years have meant a great deal to me.

To the technical reviewers who provided valuable feedback and technical accuracy at the completion of each draft.

To the many people at SAS Institute with whom I have developed so many friendships over the years. I'd like to express my thanks to each of you as well as all the knowledgeable people in SAS Technical Support. Thank you for developing and supporting such a great product and making the last 25 years a rewarding and enjoyable journey.

To Charles Edwin Shipp of Shipp Consulting and Sunil Kumar Gupta of Gupta Programming for their support and encouragement through the years.

To SAS User Groups, their leadership, and members throughout the world —you are the greatest group of professional software users anywhere.

To the many teachers I have had in my life I thank you. Special thanks go to Lawrence Delk (6th grade); Mr. Almeida (12th grade); Professor Carl Kromp (Industrial Engineering); Joseph J. Moder, PhD (Management Science); Charles N. Kurucz, PhD (Management Science); John F. Stewart, PhD (Computer Information Systems); Earl Wiener, PhD (Management Science); Howard Seth Gitlow, PhD (Management Science); Dean Paul K. Sugrue, PhD (University of Miami School of Business); Edward K. Baker III, PhD (Management Science); Robert T. Grauer, PhD (Computer Information Systems); and Ulu (Rydacom) for sharing your knowledge and enthusiasm.

To the countless people that I have worked with and the companies I have worked for — the experiences and memories have been invaluable.

To my mother, father, and brother for all the wonderful experiences and memories you have given and shared with me. Your love and encouragement through the years have fueled my desire to live life to the fullest.

Finally, to my wife Darlynn and son Ryan for your love, support, and for giving me a sense of balance between family, work, and play. I love you both so very much.

Thank you all!

~ Kirk ~

Chapter 1

Designing Database Tables

1.1 Introduction

The area of database design is very important in relational processes. Much has been written on this subject including entire textbooks and thousands of technical papers. No pretenses are made about the thoroughness of this very important subject in these pages. Rather, an attempt is made to provide a quick-start introduction for those readers unfamiliar with the issues and techniques of basic design principles. Readers needing more information are referred to the references listed in the back of this book.

1.2 Database Design

Activities related to "good" database design require the identification of end-user requirements and involve defining the structure of data values on a physical level. Database design begins with a ***conceptual view*** of what is needed. The next step, called ***logical design***, consists of developing a formal description of database entities and relationships to satisfy user requirements. Seldom does a database consist of a single table. Consequently, tables of interrelated information are created to enable more complex and powerful operations on data. In the final step, referred to as ***physical design***, the goal is to achieve optimal performance and efficient storage of the logical database.

1.2.1 Conceptual View

The health and well-being of a database depends on its database design. A database must be in balance (optimized) with all of its components to avoid performance and operation bottlenecks. Database design doesn't just happen. It involves planning, modeling, creating, monitoring, and adjusting to satisfy the endless assortment of user requirements without exhausting available resources. Of central importance to database design is the process of planning. Planning is a valuable component that, when absent, causes a database to fall prey to a host of problems including poor performance and difficulty in operation. Database design consists of three distinct phases, as illustrated below.

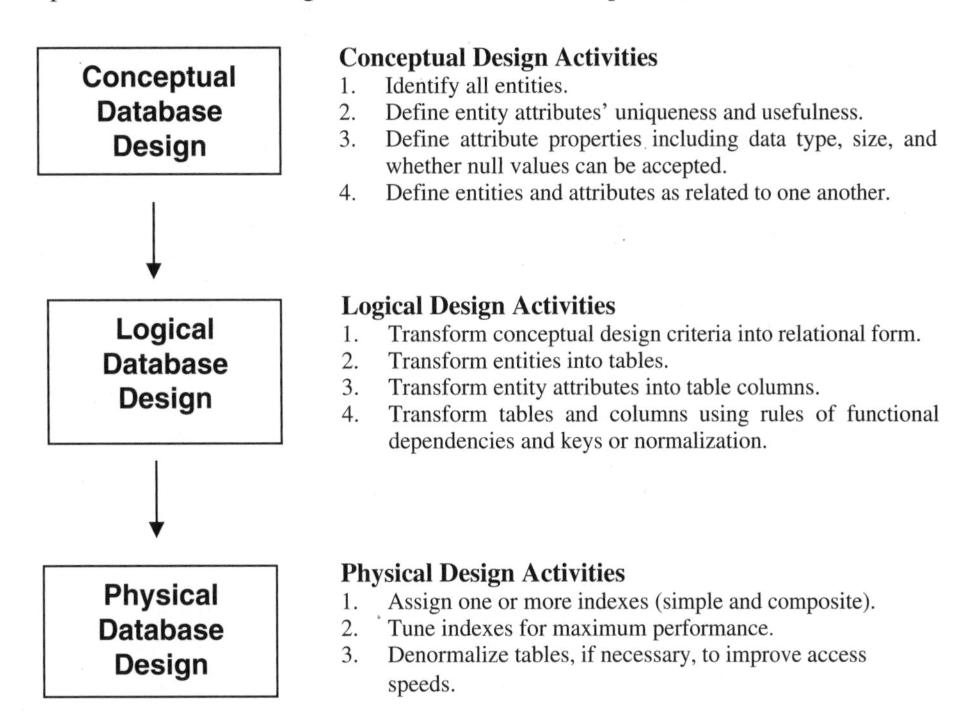

1.2.2 Table Definitions

PROC SQL uses a model of data stored as sets rather than as physical files. A physical file consists of one or more records ordered sequentially or some other way. Programming languages such as COBOL and FORTRAN evolved to process files of this type by performing operations one record at a time. These languages were generally designed and used to mimic the way people process paper forms.

PROC SQL was designed to work with sets of data. Sets have no order and members of a set are of the same type using a data structure known as a table. A table is either a base table consisting of zero or more rows with one or more columns or a virtual table called a view (see Chapter 8, "Working with Views").

1.2.3 Redundant Information

One of the rules of good database design is that data not be redundant or not be duplicated in the same database. The rationale for this is that if data appears more than once, then there is reason to believe that one of the pieces of data is likely to be in error. Another thing to watch for is the appearance of too many columns containing null values. When this occurs, the database is probably not designed properly. To alleviate potential table design issues, a process referred to as normalizing is performed. When properly done, this ensures the complete absence of redundant information in a table.

1.2.4 Normalization

Designing an optimal database design is an important element of database operations. It is also critical in achieving maximum performance and flexibility while working with tables and data. To minimize errors and duplication of data, database developers apply a concept called normalization to a logical database design.

The normalization process generally involves splitting larger multicolumn tables into two or more smaller tables containing fewer columns. The rationale for doing this is found in a set of data design guidelines called normal forms. The guidelines provide designers with a set of rules for converting one or two large database tables containing numerous columns into a normalized database consisting of multiple tables and only those columns that should be included in each table. The normalization process typically consists of no more than five steps with each succeeding step subscribing to the rules of the previous steps.

Normalizing a database helps to ensure that the database does not contain redundant information in two or more of its tables. As database designers and analysts proceed through the normalization process, many are not satisfied unless a database design is carried out to at least third normal form (3NF). Joe Celko in his popular book, *SQL for Smarties: Advanced SQL Programming* (Morgan Kaufmann, 1999), describes 3NF this way: "Informally, all the non-key columns are determined by the key, the whole key, and nothing but the key."

While the normalization guidelines are extremely useful, some database purists actually go to great lengths to remove any and all table redundancies even at the expense of performance. This is in direct contrast to other database experts who follow the guidelines less rigidly in an attempt to improve the performance of a database by only going as far as the third step (or third normal form). Whatever your preference, you should keep this in mind as you normalize database tables. A fully normalized database often requires a greater number of joins and adversely affects the speed of queries. Celko mentions that the process of joining multiple tables is costly, specifically affecting processing time and computer resources.

1.2.5 Normalization Strategies

After transforming entities and attributes from the conceptual design into a logical design, the tables and columns are created. This is when a process known as normalization occurs. Normalization refers to the process of making your database tables subscribe to certain rules. Many, if not most, database designers are satisfied when third normal form (3NF) is achieved and, for the objectives of this book, I will stop at 3NF too. To help explain the various normalization steps, an example scenario will be given.

1.2.5.1 First Normal Form (1NF)

A table is considered to be in first normal form (1NF) when all of its columns describe the table completely and when each column in a row has only one value. A table satisfies 1NF when each column in a row has a single value and no repeating group information. Essentially every table meets 1NF as long as an array, list, or other structure has not been defined. The following example illustrates a table satisfying the 1NF rule because it has only one value at each row-and-column intersection. The table is in ascending order by CUSTNUM and consists of customers and the purchases they made at an office supply store.

CUSTNUM	CUSTNAME	CUSTCITY	ITEM	UNITS	UNITCOST	MANUCITY
1	Smith	San Diego	Chair	1	$179.00	San Diego
1	Smith	San Diego	Pens	12	$0.89	Los Angeles
1	Smith	San Diego	Paper	4	$76.95	Washington
1	Smithe	San Diego	Stapler	1	$8.95	Los Angeles
7	Lafler	Spring Valley	Mouse Pad	1	$11.79	San Diego
7	Loffler	Spring Valley	Pens	24	$1.59	Los Angeles
13	Thompson	Miami	Markers	.	$0.99	Los Angeles

1.2.5.2 Second Normal Form (2NF)

The very nature of leaving a table in first normal form (1NF) may present problems because of the repetition of some information in the table as shown in the example above. Another problem is that there are misspellings in the customer names. Although repeating information may be permissible with hierarchical file structures and other legacy type file structures, it does pose a potential data consistency problem as it relates to relational data.

To describe how data consistency problems can occur, let's say that a customer takes a new job and moves to a new city. In changing the customer's city to the new location, you might find it very easy to miss one or more occurrences resulting in a customer residing incorrectly in two different cities. Assuming that our table is only meant to track one unique customer per city, this would definitely be a data consistency problem.

Essentially, second normal form (2NF) is important because it says that every nonkey column must depend on the entire primary key.

Tables that subscribe to 2NF prevent the need to make changes in more than one place. What this means in normalization terms is that tables in 2NF have no partial key dependencies. As a result, our database consisting of a single table that satisfies 1NF will need to be split into two separate tables in order to subscribe to the 2NF rule. Each table would contain the CUSTNUM column to connect the two tables. Unlike the single table in 1NF, the tables in 2NF allow a customer's city to be easily changed whenever they move to another city because the CUSTCITY column only appears once. The tables in 2NF would be constructed as follows.

CUSTOMERS Table

CUSTNUM	CUSTNAME	CUSTCITY
1	Smith	San Diego
1	Smithe	San Diego
7	Lafler	Spring Valley
13	Thompson	Miami

PURCHASES Table

CUSTNUM	ITEM	UNITS	UNITCOST	MANUCITY
1	Chair	1	$179.00	San Diego
1	Pens	12	$0.89	Los Angeles
1	Paper	4	$6.95	Washington
1	Stapler	1	$8.95	Los Angeles
7	Mouse Pad	1	$11.79	San Diego
7	Pens	24	$1.59	Los Angeles
13	Markers	.	$0.99	Los Angeles

1.2.6 Third Normal Form (3NF)

Referring to the two tables constructed according to the rules of 2NF, you may have noticed that the PURCHASES table contains a column called MANUCITY. The MANUCITY column stores the city where the product manufacturer is headquartered. Keeping this column in the PURCHASES table violates the third normal form (3NF) because MANUCITY does not provide factual information about the primary key column in the PURCHASES table. Consequently, tables are considered to be in third normal form (3NF) when each column is "dependent on the key, the whole key, and nothing but the key." The tables in 3NF are constructed so the MANUCITY column would be in a table of its own as follows.

CUSTOMERS Table

CUSTNUM	CUSTNAME	CUSTCITY
1	Smith	San Diego
1	Smithe	San Diego
7	Lafler	Spring Valley
13	Thompson	Miami

PURCHASES Table

CUSTNUM	ITEM	UNITS	UNITCOST
1	Chair	1	$179.00
1	Pens	12	$0.89
1	Paper	4	$6.95
1	Stapler	1	$8.95
7	Mouse Pad	1	$11.79
7	Pens	24	$1.59
13	Markers	.	$0.99

MANUFACTURERS Table

MANUNUM	MANUCITY
101	San Diego
112	San Diego
210	Los Angeles
212	Los Angeles
213	Los Angeles
214	Los Angeles
401	Washington

1.2.7 Beyond Third Normal Form

In general, database designers are satisfied when their database tables subscribe to the rules in 3NF. But it is not uncommon for others to normalize their database tables to fourth normal form (4NF) where independent one-to-many relationships between primary key and nonkey columns are forbidden. Some database purists will even normalize to fifth normal form (5NF) where tables are split into the smallest pieces of information in an attempt to eliminate any and all table redundancies. Although constructing tables in 5NF may provide the greatest level of database integrity, it is neither practical nor desired by most database practitioners.

There is no absolute right or wrong reason for database designers to normalize beyond 3NF as long as they have considered all the performance issues that may arise by doing so. A common problem that occurs when database tables are normalized beyond 3NF is that a large number of small tables are generated. In these cases, an increase in time and computer resources frequently occurs because small tables must first be joined before a query, report, or statistic can be produced.

1.3 Column Names and Reserved Words

The ANSI Standard reserves a number of SQL keywords from being used as column names. The SAS SQL implementation is not as rigid, but users should be aware of what reserved words exist to prevent unexpected and unintended results during SQL processing. Column names should conform to proper SAS naming conventions (as described in the *SAS Language Reference*), and they should not conflict with certain reserved words found in the SQL language. The following list identifies the reserved words found in the ANSI SQL standard.

ANSI SQL Reserved Words

AS	INNER	OUTER
CASE	INTERSECT	RIGHT
EXCEPT	JOIN	UNION
FROM	LEFT	UPPER
FULL	LOWER	USER
GROUP	ON	WHEN
HAVING	ORDER	WHERE

You probably will not encounter too many conflicts between a column name and an SQL reserved word, but when you do you will need to follow a few simple rules to prevent processing errors from occurring. As was stated earlier, although PROC SQL's naming conventions are not as rigid as other vendors' implementations, care should still be exercised, in particular when PROC SQL code is transferred to other database environments expecting it to run error-free. If a column name in an existing table conflicts with a reserved word, you have three options at your disposal:

1. Physically rename the column in the table, as well as any references to the column.
2. Use the RENAME= data set option to rename the desired column in the current query.
3. Specify the PROC SQL option DQUOTE=ANSI, and surround the column name (reserved word) in double quotes, as illustrated below.

SQL Code

```
PROC SQL DQUOTE=ANSI;
 SELECT *
  FROM RESERVED_WORDS
   WHERE "WHERE"="EXAMPLE";
QUIT;
```

1.4 Data Integrity

Webster's New World Dictionary defines integrity as "the quality or state of being complete; perfect condition; reliable; soundness." Data integrity is a critical element that every organization must promote and strive for. It is imperative that the data tables in a database environment be reliable, free of errors, and sound in every conceivable way. The existence of data errors, missing information, broken links, and other related problems in one or more tables can affect decision-making and information reporting activities resulting in a loss of confidence among users.

Applying a set of rules to the database structure and content can ensure the integrity of data resources. These rules consist of table and column constraints and will be discussed in detail in Chapter 5, "Creating, Populating, and Deleting Tables."

1.4.1 Referential Integrity

Referential integrity refers to the way in which database tables handle update and delete requests. Database tables frequently have a **primary key** where one or more columns have a unique value by which rows in a table can be identified and selected. Other tables may have one or more columns called a **foreign key** that is used to connect to some other table through its value. Database designers frequently apply rules to database tables to control what happens when a primary key value changes and its effect on one or more foreign key values in other tables. These referential integrity rules restrict the data that may be updated or deleted in tables.

Referential integrity ensures that rows in one table have corresponding rows in another table. This prevents lost linkages between data elements in one table and those of another enabling the integrity of data to always be maintained. Using the 3NF tables defined earlier, a foreign key called CUSTNUM can be defined in the PURCHASES table that corresponds to the primary key CUSTNUM column in the CUSTOMERS table. Users are referred to Chapter 5, “Creating, Populating, and Deleting Tables,” for more details on assigning referential integrity constraints.

1.5 Database Tables Used in This Book

This section describes a database or library of tables that is used by an imaginary computer hardware and software manufacturer. The library consists of six tables: customer, inventory, invoice, manufacturers, products, and purchases. The examples used throughout this book are based on this library (database) of tables and are described and displayed below. An alphabetical description of each table used throughout this book appears below.

1.5.1 CUSTOMERS Table

The CUSTOMERS table contains data on customers that have purchased computer hardware and software products from a manufacturer. Each customer is uniquely identified with a customer number. A description of each column in the customers table follows.

CUSTOMERS	
CUSTNUM	Unique number identifying the customer
CUSTNAME	Name of customer
CUSTCITY	City where customer is located

1.5.2 INVENTORY Table

The INVENTORY table contains customer inventory information consisting of computer hardware and software products. The inventory table contains no historical data. As inventories are replenished, the old quantity is overwritten with the new quantity. A description of each column in the inventory table follows.

INVENTORY	
PRODNUM	Unique number identifying product
MANUNUM	Unique number identifying the manufacturer
INVENQTY	Number of units of product in stock
ORDDATE	Date product was last ordered
INVENCST	Cost of inventory in customer's stock room

1.5.3 INVOICE Table

The INVOICE table contains information about customer purchases. Each invoice is uniquely identified with an invoice number. A description of each column in the invoice table follows:

INVOICE	
INVNUM	Unique number identifying the invoice
MANUNUM	Unique number identifying the manufacturer
CUSTNUM	Customer number
PRODNUM	Product number
INVQTY	Number of units sold
INVPRICE	Unit price

1.5.4 MANUFACTURERS Table

The MANUFACTURERS table contains data about companies that make computer hardware and software products. Two companies cannot have the same name. No historical data is kept in this table. If a company is sold or stops making computer hardware or software, then the manufacturer is dropped from the table. In the event a manufacturer has an address change, the old address is overwritten with the new address. A description of each column in the manufacturers table follows.

MANUFACTURERS	
MANUNUM	Unique number identifying the manufacturer
MANUNAME	Name of manufacturer
MANUCITY	City where manufacturer is located
MANUSTAT	State where manufacturer is located

1.5.5 PRODUCTS Table

The PRODUCTS table contains data about computer hardware and software products offered for sale by the manufacturer. Each product is uniquely identified with a product number. A description of each column in the products table follows.

PRODUCTS	
PRODNUM	Unique number identifying the product
PRODNAME	Name of product
MANUNUM	Unique number identifying the manufacturer
PRODTYPE	Type of product
PRODCOST	Cost of product

1.5.6 PURCHASES Table

The PURCHASES table contains information about computer hardware and software products purchased by customers. Each product is uniquely identified with a product number. A description of each column in the purchases table follows.

PURCHASES

CUSTNUM	Unique number identifying the product
ITEM	Name of product
UNITS	Unique number identifying the manufacturer
UNITCOST	Cost of product

1.6 Table Contents

An alphabetical list of tables, variables, and attributes for each table is displayed below.

Customers CONTENTS Output

-----Alphabetic List of Variables and Attributes-----

Variable	Type	Len	Pos	Label
custcity	Char	20	25	Customer's Home City
custname	Char	25	0	Customer Name
custnum	Num	3	45	Customer Number

Inventory CONTENTS Output

-----Alphabetic List of Variables and Attributes-----

Variable	Type	Len	Pos	Format	Informat	Label
invencst	Num	6	10	DOLLAR10.2		Inventory Cost
invenqty	Num	3	7			Inventory Quantity
manunum	Num	3	16			Manufacturer Number
orddate	Num	4	0	MMDDYY10.	MMDDYY10	Date Inventory Last Ordered
prodnum	Num	3	4			Product Number

Invoice CONTENTS Output

-----Alphabetic List of Variables and Attributes-----

Variable	Type	Len	Pos	Format	Label
custnum	Num	3	6		Customer Number
invnum	Num	3	0		Invoice Number
invprice	Num	5	12	DOLLAR12.2	Invoice Unit Price
invqty	Num	3	9		Invoice Quantity - Units Sold
manunum	Num	3	3		Manufacturer Number
prodnum	Num	3	17		Product Number

Manufacturers CONTENTS Output

-----Alphabetic List of Variables and Attributes-----

Variable	Type	Len	Pos	Label
manucity	Char	20	25	Manufacturer City
manuname	Char	25	0	Manufacturer Name
manunum	Num	3	47	Manufacturer Number
manustat	Char	2	45	Manufacturer State

Products CONTENTS Output

-----Alphabetic List of Variables and Attributes-----

Variable	Type	Len	Pos	Format	Label
manunum	Num	3	40		Manufacturer Number
prodcost	Num	5	43	DOLLAR9.2	Product Cost
prodname	Char	25	0		Product Name
prodnum	Num	3	48		Product Number
prodtype	Char	15	25		Product Type

Purchases CONTENTS Output

-----Alphabetic List of Variables and Attributes-----

Variable	Type	Len	Pos	Format
custnum	Num	4	0	
item	Char	10	8	
unitcost	Num	4	4	DOLLAR12.2
units	Num	3	18	

1.6.1 The Database Structure

The logical relationship between each table and the columns common to each appear below.

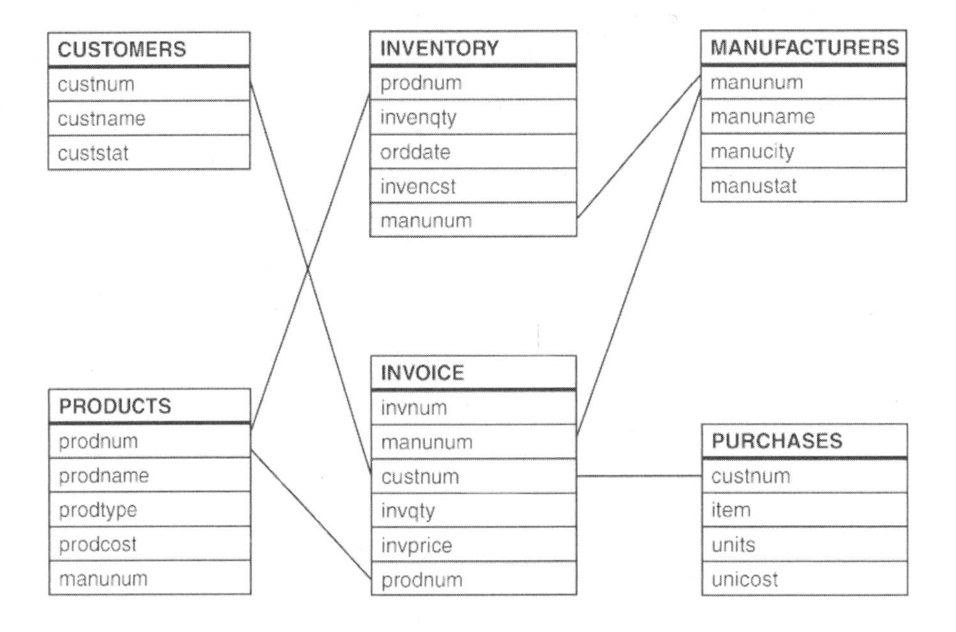

1.6.2 Sample Database Tables

The six tables (named above) represent a relational database that will be illustrated in the examples in this book. These tables are small enough to follow easily, but complex enough to illustrate the power of SQL. The data contained in each table appears below.

CUSTOMERS Table

custnum	custname	custcity
101	La Mesa Computer Land	La Mesa
201	Vista Tech Center	Vista
301	Coronado Internet Zone	Coronado
401	La Jolla Computing	La Jolla
501	Alpine Technical Center	Alpine
601	Oceanside Computer Land	Oceanside
701	San Diego Byte Store	San Diego
801	Jamul Hardware & Software	Jamul
901	Del Mar Tech Center	Del Mar
1001	Lakeside Software Center	Lakeside
1101	Bonsall Network Store	Bonsall
1201	Rancho Santa Fe Tech	Rancho Santa Fe
1301	Spring Valley Byte Center	Spring Valley
1401	Poway Central	Poway
1501	Valley Center Tech Center	Valley Center
1601	Fairbanks Tech USA	Fairbanks Ranch

(*continued on next page*)

custnum	custname	custcity
1701	Blossom Valley Tech	Blossom Valley
1801	Chula Vista Networks	

N = 18

INVENTORY Table

prodnum	invenqty	orddate	invencst	manunum
1110	20	09/01/2000	$45,000.00	111
1700	10	08/15/2000	$28,000.00	170
5001	5	08/15/2000	$1,000.00	500
5002	3	08/15/2000	$900.00	500
5003	10	08/15/2000	$2,000.00	500
5004	20	09/01/2000	$1,400.00	500
5001	2	09/01/2000	$1,200.00	600

N = 7

INVOICE Table

invnum	manunum	custnum	invqty	invprice	prodnum
1001	500	201	5	$1,495.00	5001
1002	600	1301	2	$1,598.00	6001
1003	210	101	7	$245.00	2101
1004	111	501	3	$9,600.00	1110
1005	500	801	2	$798.00	5002
1006	500	901	4	$396.00	6000
1007	500	401	7	$23,100.00	1200

N = 7

MANUFACTURERS Table

manunum	manuname	manucity	manustat
111	Cupid Computer	Houston	TX
210	KPL Enterprises	San Diego	CA
600	World Internet Corp	Miami	FL
120	Storage Devices Inc	San Mateo	CA
500	Global Software	San Diego	CA
700	San Diego PC Planet	San Diego	CA

N = 6

PRODUCTS Table

prodnum	prodname	manunum	prodtype	prodcost
1110	Dream Machine	111	Workstation	$3,200.00
1200	Business Machine	120	Workstation	$3,300.00
1700	Travel Laptop	170	Laptop	$3,400.00
2101	Analog Cell Phone	210	Phone	$35.00
2102	Digital Cell Phone	210	Phone	$175.00
2200	Office Phone	220	Phone	$130.00
5001	Spreadsheet Software	500	Software	$299.00
5002	Database Software	500	Software	$399.00
5003	Wordprocessor Software	500	Software	$299.00
5004	Graphics Software	500	Software	$299.00

N=10

PURCHASES Table

custnum	item	units	unitcost
1	Chair	1	$179.00
1	Pens	12	$0.89
1	Paper	4	$6.95
1	Stapler	1	$8.95
7	Mouse Pad	1	$11.79

(continued on next page)

custnum	item	units	unitcost
7	Pens	24	$1.59
13	Markers	.	$0.99
N=7			

1.7 Summary

1. Poor database design is often attributed to the relative ease by which tables can be created and populated in a relational database. By adhering to certain rules, good design can be structured into almost any database (see section 1.2.1).

2. SQL was designed to work with sets of data and accesses a data structure known as a table (see section 1.2.2).

3. Achieving optimal design of a database means that the database contains little or no redundant information in two or more of its tables. This means that good database design calls for little or no replication of data (see section 1.2.3).

4. Poor database design can result in costly or inefficient processing, coding complexities, complex logical relationships, long application development times, or excessive storage requirements (see section 1.2.4).

5. Design decisions made in one phase may involve making one or more tradeoffs in another phase (see section 1.2.4).

6. A database in third normal form (3NF) is where a column is "dependent on the key, the whole key, and nothing but the key" (see section 1.2.4).

Chapter 2

Working with Data in PROC SQL

2.1 Introduction

PROC SQL is essentially a database language as opposed to a procedural or computational language. Although only two data types are available in the SAS System's implementation of SQL, numerous extensions including operators, functions, and predicates are available to PROC SQL programmers.

2.2 Data Types Overview

The purpose of a database is to store data. A database contains one or more tables (and other components). Tables consist of columns and rows of data. In the SAS implementation of SQL, there are two available data types: 1) numeric and 2) character.

2.2.1 Numeric Data

The SAS implementation of SQL provides programmers with numerous arithmetic, statistical, and summary functions. It offers one numeric data type to represent numeric data. Columns defined as a numeric data type with the NUMERIC or NUM column definition are assigned a default length of 8 bytes, even if the column is created with a numeric length less than 8 bytes. This provides the greatest degree of precision allowed by the SAS System. In the example below, a table called PURCHASES is created consisting of two character and two numeric columns. The resulting table contains no rows of data, as illustrated by the SAS log results. For more information about the CREATE TABLE statement, see Chapter 5, "Creating, Populating, and Deleting Tables."

SQL Code

```
PROC SQL;
  CREATE TABLE PURCHASES
    (CUSTNUM  CHAR(4),
     ITEM     CHAR(10),
     UNITS    NUM
     UNITCOST NUM(8,2));
QUIT;
```

SAS Log Results

```
     PROC SQL;
       CREATE TABLE PURCHASES
         (CUSTNUM  CHAR(4),
          ITEM     CHAR(10),
          UNITS    NUM,
          UNITCOST NUM(8,2));
NOTE: Table PURCHASES created, with 0 rows and 4
columns.
     QUIT;
```

Results

```
                The CONTENTS Procedure

-----Alphabetic List of Variables and Attributes-----

          #    Variable     Type    Len    Pos
          ___________________________________
          1    CUSTNUM      Char      4     16
          2    ITEM         Char     10     20
          3    UNITS        Num       8      0
          4    UNITCOST     Num       8      8
```

Use the DATA step LENGTH statement to create a column of fewer than 8 bytes. Although this action can cause precision issues, it allows for more efficient use of data storage resources. See the *SAS Language Reference: Dictionary* for more information. The example illustrates a DATA step that assigns smaller lengths to the two numeric variables, UNITS and UNITCOST, in the PURCHASES table. The CONTENTS output illustrates the creation of shorter length numeric variables.

DATA Step Code

```
DATA PURCHASES;
  LENGTH CUSTNUM  $4.
         ITEM    $10.
         UNITS     3.
         UNITCOST  4.;
  LABEL CUSTNUM  = 'Customer Number'
        ITEM     = 'Item Purchased'
        UNITS    = '# Units Purchased'
        UNITCOST = 'Unit Cost';
  FORMAT UNITCOST DOLLAR12.2;
RUN;
PROC CONTENTS DATA=PURCHASES;
RUN;
```

SAS Log Results

```
DATA PURCHASES;
  LENGTH CUSTNUM  $4.
         ITEM    $10.
         UNITS     3.
         UNITCOST  4.;
  LABEL CUSTNUM  = 'Customer Number'
         ITEM     = 'Item Purchased'
         UNITS    = '# Units Purchased'
         UNITCOST = 'Unit Cost';
  FORMAT UNITCOST DOLLAR12.2;
RUN;

NOTE: Variable CUSTNUM is uninitialized.
NOTE: Variable ITEM is uninitialized.
NOTE: Variable UNITS is uninitialized.
NOTE: Variable UNITCOST is uninitialized.
NOTE: The data set WORK.PURCHASES has 1 observations
      and 4 variables.
NOTE: DATA statement used:
      real time           2.80 seconds

PROC CONTENTS DATA=PURCHASES;
RUN;

NOTE: PROCEDURE CONTENTS used:
      real time           1.82 seconds
```

CONTENTS Results

```
                   The CONTENTS Procedure

      -----Alphabetic List of Variables and Attributes-----

 #    Variable    Type    Len    Pos    Format        Label
 ________________________________________________________________
 1    CUSTNUM     Char      4      4                  Customer Number
 2    ITEM        Char     10      8                  Item Purchased
 3    UNITS       Num       3     18                  # Units Purchased
 4    UNITCOST    Num       4      0    DOLLAR12.2    Unit Cost
```

2.2.2 Date and Time Column Definitions

Database application processing stores date and time information in the form of a numeric data type. Date and time values are represented internally as an offset where a SAS date value is stored as the number of days from the fixed date value of 01/01/1960 (January 1, 1960). The SAS date value for January 1, 1960, is represented as 0 (zero). A date earlier than this is represented as a negative number, and a date later than this is represented as a positive number. This makes performing date calculations much easier.

The SAS System has integrated the vast array of date and time informats and formats with PROC SQL. The various informats and formats act as input and output templates and describe how date and time information is to be read or rendered on output. See the *SAS Language Reference*: *Dictionary* for detailed descriptions of the various informats and formats and their use. Numeric date and time columns, when combined with informats and/or formats, automatically validate values according to the following rules:

Date Date informats and formats enable PROC SQL and the SAS System to determine the month, day, and year values of a date. The month value handles values between 1 through 12. The day value handles values from 1 through 31 and applies additional validations to a maximum of 28, 29, or 30 depending on the month in question. The year value handles values 1 through 9999. Dates go back to 1582 and ahead to 20,000. When you enter a year value of 0001 and specify a format and yearcutoff value of 1920, the returned value would be 2001.

Time	Time informats and formats enable PROC SQL to determine the hour, minute, and second values of a time. The hour portion handles values between 00 and 23. The minute portion handles values from 00 through 59. The second portion handles values from 00 through 59.
DATETIME	Date and time stamps enable PROC SQL to determine the month, day, and year of a date as well the hour, minute, and second of a time.

See Chapter 5, "Creating, Populating, and Deleting Tables" and Chapter 6, "Modifying and Updating Tables and Indexes," for more information about date and time informats and formats.

2.2.3 Character Data

PROC SQL provides tools to manipulate and store character data including words, text, and codes using the CHARACTER or CHAR data type. The characters allowed by this data type include the ASCII or EBCDIC character sets. The CHARACTER or CHAR data type stores fixed-length character strings consisting of a maximum of 32K characters. If a length is not specified, a CHAR column stores a default of 8 characters.

The SQL programmer has a vast array of SQL and Base SAS functions that can make the task of working with character data considerably easier. In this chapter you'll learn how columns based on the character data type are defined and how string functions, pattern matching, phonetic matching techniques, and a variety of other techniques are used with character data.

2.2.4 Missing Values and Null

Missing values are an important aspect of dealing with data. The concept of missing values is familiar to programmers, statisticians, researchers, and other SAS users. This section describes what null values are, what they aren't, and how they are used.

Missing or unknown information is supported by PROC SQL in a form known as a null value. A null value is not the same as a zero value. In the SAS System, nulls are treated as a separate category from known values. A value consisting of zero has a known value. In contrast, a value of null has an unknown quantity and will never be known. For example, a patient who is given an eye exam does not have zero eyesight just because the results from the exam haven't been received. The correct value to assign in a case like this is a missing or a null value.

In another example, say a person declines to provide their age on a survey. This person's age is null, not zero. Essentially, this person has an age, but it is unknown. Whenever an unknown value occurs, you have no choice but to assign an unknown value – null.

Since the value of null is unknown, any arithmetic calculation using a null will return a null. This makes a lot of sense since the results of a calculation using a null are not determinable. This is sometimes referred to as the propagation of nulls because when a null value is used in a calculation or an expression, it propagates a null value. For example, if a null is added to a known value, the result is a null value.

2.2.5 Arithmetic and Missing Data

In the SAS System, a numeric data type containing a null value (absence of any value) is represented with a period (.) This representation indicates that the column has not been assigned a value. A null value has no value and is not the same as zero. A value consisting of zero has a known quantity as opposed to a null value that is not known and never will be known.

If a null value is multiplied with a known value, the result is a null value represented with a period (.). In the next example, when UNITS and UNITCOST both have known values, their product will generate a known value, as is illustrated for the Markers purchase below.

SQL Code

```
PROC SQL;
  SELECT CUSTNUM,
         ITEM,
         UNITS,
         UNITCOST,
         UNITS * UNITCOST AS TOTAL
    FROM PURCHASES
      ORDER BY TOTAL;
QUIT;
```

Results

```
                    The SAS System

custnum  item            units      unitcost       TOTAL

     13  Markers             .         $0.99           .
      1  Stapler             1         $8.95    8.949997
      1  Pens               12         $0.89       10.68
      7  Mouse Pad           1        $11.79    11.78999
      1  Paper               4         $6.95    27.79999
      7  Pens               24         $1.59    38.15998
      1  Chair               1       $179.00         179
```

2.2.6 SQL Keywords

SQL provides three keywords: AS, DISTINCT, and UNIQUE to perform specific operations on the results. Each will be presented in order as follows.

2.2.6.1 Creating Column Aliases

In situations where data is computed using system functions, statistical functions, or arithmetic operations, a column name or header can be left blank. To prevent this from occurring, users may specify the **AS** keyword to provide a name to the column or heading itself. The next example illustrates using the AS keyword to prevent the name for the computed column from being assigned a temporary column name similar to: _TEMAxxx. The name assigned with the AS keyword is also used as the column header on output, as shown.

SQL Code

```
PROC SQL;
  SELECT PRODNAME,
         PRODTYPE,
         PRODCOST * 0.80  AS Discount_Price
    FROM PRODUCTS
      ORDER BY 3;
QUIT;
```

Results

```
                         The SAS System

   Product Name                Product Type     Discount_Price
   ---------------------------------------------------------
   Analog Cell Phone           Phone                        28
   Office Phone                Phone                       104
   Digital Cell Phone          Phone                       140
   Spreadsheet Software        Software                  239.2
   Graphics Software           Software                  239.2
   Wordprocessor Software      Software                  239.2
   Database Software           Software                  319.2
   Dream Machine               Workstation                2560
   Business Machine            Workstation                2640
   Travel Laptop               Laptop                     2720
```

2.2.6.2 Finding Duplicate Values

In some situations, several rows in a table will contain identical column values. To select only one of each duplicate value, the **DISTINCT** keyword can be used in the SELECT statement as follows.

SQL Code

```
PROC SQL;
  SELECT DISTINCT MANUNUM
    FROM INVENTORY;
QUIT;
```

Results

```
                     The SAS System

                     Manufacturer
                           Number
                     ------------
                              111
                              170
                              500
                              600
```

2.2.6.3 Finding Unique Values

In some situations, several rows in a table will contain identical column values. To select each of these duplicate values only once, the **UNIQUE** keyword can be used in the SELECT statement.

SQL Code

```
PROC SQL;
  SELECT UNIQUE MANUNUM
    FROM INVENTORY;
QUIT;
```

Results

```
                The SAS System

               Manufacturer
                     Number
               ------------
                        111
                        170
                        500
                        600
```

2.3 SQL Operators and Functions

SQL programmers have a number of ways to accomplish their objectives, particularly when the goal is to retrieve and work with data. The SELECT statement is an extremely powerful statement in the SQL language. Its syntax can be somewhat complex because of the number of ways that columns, tables, operators, functions, and predicates can be combined into executable statements.

There are several types of operators and functions in PROC SQL: 1) comparison operators, 2) logical operators, 3) arithmetic operators, 4) character string operators, 5) summary functions, and 6) predicates. Operators and functions provide value-added features for PROC SQL programmers. Each will be presented below.

2.3.1 Comparison Operators

Comparison operators are used in the SQL procedure to compare one character or numeric value to another. As in the DATA step, SQL comparison operators, mnemonics, and their descriptions appear in the following table.

SAS Operator	Mnemonic Operator	Description
=	EQ	Equal to
^= or ¬=	NE	Not equal to
<	LT	Less than
<=	LE	Less than or equal to
>	GT	Greater than
>=	GE	Greater than or equal to

Suppose you want to select only those products from the PRODUCTS table costing more than $300.00. The example below illustrates the use of the greater than sign (>) in a WHERE clause to select products meeting the condition.

SQL Code

```
PROC SQL;
  SELECT PRODNAME,
         PRODTYPE,
         PRODCOST
    FROM PRODUCTS
      WHERE PRODCOST > 300;
QUIT;
```

Results

```
                    The SAS System

                                                    Product
Product Name                  Product Type             Cost
-----------------------------------------------------------
Dream Machine                 Workstation         $3,200.00
Business Machine              Workstation         $3,300.00
Travel Laptop                 Laptop              $3,400.00
Database Software             Software              $399.00
```

PROC SQL also supports the use of truncated string comparison operators. These operators work by first truncating the longer string to the same length as the shorter string, and then performing the specified comparison. Using any of the comparison operators has no permanent effect on the strings themselves. The list of truncated string comparison operators and their meanings appears below.

Truncated String Comparison Operator	Description
EQT	Equal to
GTT	Greater than
LTT	Less than
GET	Greater than or equal to
LET	Less than or equal to
NET	Not equal to

2.3.2 Logical Operators

Logical operators are used to connect two or more expressions together in a WHERE or HAVING clause. The available logical operators are AND, OR, and NOT. Suppose you want to select only those software products costing more than $300.00. The example illustrates how the AND operator is used to ensure that both conditions are true.

SQL Code

```
PROC SQL;
  SELECT PRODNAME,
         PRODTYPE,
         PRODCOST
    FROM PRODUCTS
      WHERE PRODTYPE = 'Software' AND
            PRODCOST > 300;
QUIT;
```

Results

The SAS System

Product Name	Product Type	Product Cost
Database Software	Software	$399.00

The next example illustrates the use of the OR logical operator to select software products or products that cost more than $300.

SQL Code

```
PROC SQL;
  SELECT PRODNAME,
         PRODTYPE,
         PRODCOST
    FROM PRODUCTS
      WHERE PRODTYPE = 'Software' OR
            PRODCOST > 300;
QUIT;
```

Results

The SAS System

Product Name	Product Type	Product Cost
Dream Machine	Workstation	$3,200.00
Business Machine	Workstation	$3,300.00
Travel Laptop	Laptop	$3,400.00
Spreadsheet Software	Software	$299.00
Database Software	Software	$399.00
Wordprocessor Software	Software	$299.00
Graphics Software	Software	$299.00

The next example illustrates the use of the NOT logical operator to select products that are not software products and do not cost more than $300.

SQL Code

```
PROC SQL;
  SELECT PRODNAME,
         PRODTYPE,
         PRODCOST
    FROM PRODUCTS
      WHERE NOT PRODTYPE = 'Software' AND
            NOT PRODCOST > 300;
QUIT;
```

Results

```
                          The SAS System

                                                           Product
 Product Name                   Product Type                  Cost
 ------------------------------------------------------------------
 Analog Cell Phone              Phone                       $35.00
 Digital Cell Phone             Phone                      $175.00
 Office Phone                   Phone                      $130.00
```

2.3.3 Arithmetic Operators

The arithmetic operators used in PROC SQL are the same as those used in the DATA step as well as those found in other languages like C, Pascal, FORTRAN, and COBOL. The arithmetic operators available in the SQL procedure appear below.

Operator	Description
+	Addition
-	Subtraction
*	Multiplication
/	Division
**	Exponent (raises to a power)
=	Equals

To illustrate how arithmetic operators are used, suppose you want to apply a discount of 20% to the product price (PRODCOST) in the PRODUCTS table and display the results in ascending order by the discounted price. Note that the computed column (PRODCOST * 0.80) does not automatically create a column header on output.

SQL Code

```
PROC SQL;
  SELECT PRODNAME,
         PRODTYPE,
         PRODCOST * 0.80
    FROM PRODUCTS;
QUIT;
```

Results

The SAS System

Product Name	Product Type	
Dream Machine	Workstation	2560
Business Machine	Workstation	2640
Travel Laptop	Laptop	2720
Analog Cell Phone	Phone	28
Digital Cell Phone	Phone	140
Office Phone	Phone	104
Spreadsheet Software	Software	239.2
Database Software	Software	319.2
Wordprocessor Software	Software	239.2
Graphics Software	Software	239.2

In the next example, suppose you wanted to reference a column that was calculated in the SELECT statement. PROC SQL allows references to a computed column in the same SELECT statement (or a WHERE clause) using the CALCULATED keyword. Note that the computed columns have column aliases created for them using the AS keyword. If the CALCULATED keyword were not specified preceding the calculated column, an error would have been generated.

SQL Code

```
PROC SQL;
  SELECT PRODNAME,
         PRODTYPE,
         PRODCOST * 0.80 AS DISCOUNT_PRICE
                  FORMAT=DOLLAR9.2,
         PRODCOST - CALCULATED DISCOUNT_PRICE AS LOSS
                  FORMAT=DOLLAR7.2
    FROM PRODUCTS
      ORDER BY 3;
QUIT;
```

Results

The SAS System

Product Name	Product Type	DISCOUNT_ PRICE	LOSS
Analog Cell Phone	Phone	$28.00	$7.00
Office Phone	Phone	$104.00	$26.00
Digital Cell Phone	Phone	$140.00	$35.00
Spreadsheet Software	Software	$239.20	$59.80
Graphics Software	Software	$239.20	$59.80
Wordprocessor Software	Software	$239.20	$59.80
Database Software	Software	$319.20	$79.80
Dream Machine	Workstation	$2,560.00	$640.00
Business Machine	Workstation	$2,640.00	$660.00
Travel Laptop	Laptop	$2,720.00	$680.00

2.3.4 Character String Operators and Functions

Character string operators and functions are typically used with character data. Numerous operators are presented to make you aware of the power available with the SQL procedure. As you become familiar with each of the operators, you'll find their real strength as you begin to nest functions within each other.

2.3.4.1 Concatenating Strings Together

The following example illustrates a basic concatenation operator that is used to concatenate two columns and a text string. Note that the created column is not labeled. The concatenation operator will be discussed in greater detail in Chapter 3, "Formatting Output."

SQL Code

```
PROC SQL;
  SELECT MANUCITY || "," || MANUSTAT
    FROM MANUFACTURERS;
QUIT;
```

```
                 The SAS System

          ______________________________
          Houston                    ,TX
          San Diego                  ,CA
          Miami                      ,FL
          San Mateo                  ,CA
          San Diego                  ,CA
          San Diego                  ,CA
```

2.3.4.2 Finding the Length of a String

The LENGTH function is used to obtain the length of a character string column. LENGTH returns a number equal to the number of characters in the argument. Note that the computed column (LENGTH(PRODNAME)) has a column header created for it called Length by specifying the AS keyword. This example illustrates using the LENGTH function to determine the length of data values.

SQL Code

```
PROC SQL;
  SELECT PRODNUM,
         PRODNAME,
         LENGTH(PRODNAME) AS Length
    FROM PRODUCTS;
QUIT;
```

Results

```
                    The SAS System

  Product
   Number  Product Name                        Length
  ---------------------------------------------------
     1110  Dream Machine                           13
     1200  Business Machine                        16
     1700  Travel Laptop                           13
     2101  Analog Cell Phone                       17
     2102  Digital Cell Phone                      18
     2200  Office Phone                            12
     5001  Spreadsheet Software                    20
     5002  Database Software                       17
     5003  Wordprocessor Software                  22
     5004  Graphics Software                       17
```

2.3.4.3 Combining Functions and Operators

As in the DATA step, many functions can be used in the SQL procedure. To modify one or more existing rows in a table, the UPDATE statement is used (see Chapter 6, "Modifying and Updating Tables and Indexes," for more details). The UPDATE statement with SET clause changes the contents of a data value (functioning the same way as a DATA step assignment statement) by assigning a new value to the column identified to the left of the equal sign by a constant or expression referenced to the right of the equal sign.

The UPDATE statement does not automatically produce any output, except for the log messages based on the operation results itself. To illustrate the use of DATA step functions and operators in the SQL procedure, the next example shows a SCAN function that isolates the first piece of information from product name (PRODNAME), a TRIM function to remove trailing blanks from product type (PRODTYPE), and a concatenation operator "||" that concatenates the first character expression with the second expression. Exercise care when using the SCAN function because it returns a 200-byte string.

SQL Code

```
PROC SQL;
  UPDATE PRODUCTS
    SET PRODNAME = SCAN(PRODNAME,1) || TRIM(PRODTYPE);
QUIT;
```

SAS Log Results

```
      PROC SQL;
        UPDATE PRODUCTS
          SET PRODNAME = SCAN(PRODNAME,1) || TRIM(PRODTYPE);
NOTE: 10 rows were updated in PRODUCTS.
      QUIT;
```

An optional WHERE clause can be specified limiting the number of rows that modifications will be applied to. The next example illustrates using a WHERE clause to restrict the number of rows that are updated in the previous example to just "phones", excluding all the other rows.

SQL Code

```
PROC SQL;
  UPDATE PRODUCTS
    SET PRODNAME = SCAN(PRODNAME,1) || TRIM(PRODTYPE)
      WHERE PRODTYPE IN ('Phone');
QUIT;
```

SAS Log Results

```
     PROC SQL;
       UPDATE PRODUCTS
         SET PRODNAME = SCAN(PRODNAME,1) || TRIM(PRODTYPE)
           WHERE PRODTYPE IN ('Phone');
NOTE: 3 rows were updated in PRODUCTS.
     QUIT;
```

2.3.4.4 Aligning Characters

The default alignment for character data is to the left. However, character columns or expressions can also be aligned to the right. Two functions are available for character alignment: LEFT and RIGHT. The next example combines the concatenation operator "||" and the TRIM function with the **LEFT** function to left align a character expression while inserting a comma "," and blank between the columns.

SQL Code

```
PROC SQL;
  SELECT LEFT(TRIM(MANUCITY) || ", " || MANUSTAT)
    FROM MANUFACTURERS;
QUIT;
```

Results

```
                    The SAS System


                 ________________________
                 Houston, TX
                 San Diego, CA
                 Miami, FL
                 San Mateo, CA
                 San Diego, CA
                 San Diego, CA
```

The next example illustrates how character data can be right aligned using the **RIGHT** function.

SQL Code

```
PROC SQL;
  SELECT RIGHT(MANUCITY)
    FROM MANUFACTURERS;
QUIT;
```

Results

```
                    The SAS System


                  ____________________
                               Houston
                             San Diego
                                 Miami
                             San Mateo
                             San Diego
                             San Diego
```

2.3.4.5 Finding the Occurrence of a Pattern with INDEX

To find the occurrence of a pattern, the **INDEX** function can be used. Frequently, requirements call for a column to be searched using a specific character string. The INDEX function can be used in the SQL procedure to search for patterns in a character string. The character string is searched from left to right for the first occurrence of the specified value. If the desired string is found, the column position of the first character is returned. Otherwise a value of zero (0) is returned. The following arguments are used to search for patterns in a column: the character column or expression and the character string to search for. To find all products with the characters “phone” in the product name (PRODNAME) column, the following code can be specified:

SQL Code

```
PROC SQL;
  SELECT PRODNUM,
         PRODNAME,
         PRODTYPE
    FROM PRODUCTS
      WHERE INDEX(PRODNAME, 'phone') > 0;
QUIT;
```

Results

```
     PROC SQL;
       SELECT PRODNUM,
              PRODNAME,
              PRODTYPE
         FROM PRODUCTS
           WHERE INDEX(PRODNAME, 'phone') > 0;
NOTE: No rows were selected.
     QUIT;
```

2.3.4.6 Analysis

As in the DATA step, no rows were selected because the search is case-sensitive and "phone" is specified as all lowercase characters.

2.3.4.7 Changing the Case in a String

The SAS System provides two functions that enable you to change the case of a string's characters: LOWCASE and UPCASE. The LOWCASE function converts all of the characters in a string or expression to lowercase characters. The UPCASE function converts all of the characters in a string or expression to uppercase characters.

In the previous example, the results of the search were negative even though the character string "phone" appeared multiple times in more than one row. In order to make this search recognize all the possible lower- and uppercase variations of the word "phone", the search criteria in the WHERE clause could be made "smarter" by combining an **UPCASE** function with the INDEX function as follows.

SQL Code

```
PROC SQL;
  SELECT PRODNUM,
         PRODNAME,
         PRODTYPE
    FROM PRODUCTS
      WHERE INDEX(UPCASE(PRODNAME), 'PHONE') > 0;
QUIT;
```

Results

```
                    The SAS System

 Product
  Number  Product Name                        Product Type
 ----------------------------------------------------------
    2101  Analog Cell Phone                   Phone
    2102  Digital Cell Phone                  Phone
    2200  Office Phone                        Phone
```

In the next example, the **LOWCASE** function is combined with the INDEX function to produce the identical output from the previous example.

SQL Code

```
PROC SQL;
  SELECT PRODNUM,
         PRODNAME,
         PRODTYPE
    FROM PRODUCTS
      WHERE INDEX(LOWCASE(PRODNAME), 'phone') > 0;
QUIT;
```

Results

```
                      The SAS System

 Product
  Number  Product Name                  Product Type
 ------------------------------------------------------
    2101  Analog Cell Phone             Phone
    2102  Digital Cell Phone            Phone
    2200  Office Phone                  Phone
```

2.3.4.8 Extracting Information from a String

Occasionally, processing requirements call for specific pieces of information to be extracted from a column. In these situations the SUBSTR function can be used with a character column by specifying a starting position and the number of characters to extract. The following example illustrates how the **SUBSTR** function is used to capture the first 4 bytes from the product type (PRODTYPE) column.

SQL Code

```
PROC SQL;
  SELECT PRODNUM,
         PRODNAME,
         PRODTYPE,
         SUBSTR(PRODTYPE,1,4)
    FROM PRODUCTS
      WHERE PRODCOST > 100.00;
QUIT;
```

Results

```
                           The SAS System

 Product
  Number  Product Name                  Product Type

    1110  Dream Machine                 Workstation       Work
    1200  Business Machine              Workstation       Work
    1700  Travel Laptop                 Laptop            Lapt
    2102  Digital Cell Phone            Phone             Phon
    2200  Office Phone                  Phone             Phon
    5001  Spreadsheet Software          Software          Soft
    5002  Database Software             Software          Soft
    5003  Wordprocessor Software        Software          Soft
    5004  Graphics Software             Software          Soft
```

2.3.4.9 Phonetic Matching (Sounds-Like Operator =*)

A technique for finding names that sound alike or have spelling variations is available in the SQL procedure. This frequently used technique is referred to as phonetic matching and is performed using the Soundex algorithm. In Joe Celko's book, *SQL for Smarties: Advanced SQL Programming* (pages 83-87), he traced the origins of the Soundex algorithm to the developers Margaret O'Dell and Robert C. Russell in 1918. Developed before the first computer, this technique was often used by clerks to manually search for similar sounding names.

Although not technically a function, the sounds-like operator "=*" searches and selects character data based on two expressions: the search value and the matched value. Anyone that has looked for a last name in a local telephone directory is quickly reminded of the possible phonetic variations.

To illustrate how the sounds-like operator works, we will search on last name in a table called CUSTOMERS2. The CUSTOMERS2 table is illustrated below. Although each name has phonetic variations and sounds the same, the results of "Laughler," "Loffler," and "Laffler" are spelled differently (illustrated below). The following PROC SQL code uses the sounds-like operator to find all customers that sound like "Lafler".

CUSTOMERS2 Table

CUSTNUM	CUSTNAME	CUSTCITY
1	Smith	San Diego
7	Lafler	Spring Valley
11	Jones	Carmel
13	Thompson	Miami
7	Loffler	Spring Valley
1	Smithe	San Diego
7	Laughler	Spring Valley
7	Laffler	Spring Valley

SQL Code

```
PROC SQL;
  SELECT CUSTNUM,
         CUSTNAME,
         CUSTCITY
    FROM CUSTOMERS2
      WHERE CUSTNAME =* 'Lafler';
QUIT;
```

Results

The SAS System

CUSTNUM	CUSTNAME	CUSTCITY
7	Lafler	Spring Valley
7	Loffler	Spring Valley
7	Laffler	Spring Valley

Readers familiar with the DATA step SOUNDEX(argument) function to search a string are cautioned that it cannot be used in an SQL WHERE clause. Instead the sounds-like operator “=*” must be specified; otherwise a result of no rows will be selected.

Notice that only three of the four possible phonetic matches were selected in the preceding example (that is, Lafler, Loffler, and Laffler). The fourth possibility “Laughler” was not chosen as a “matched” value in the search by the sounds-like algorithm. In an attempt to overcome the inherent limitation with the sounds-like operator, as described in Celko’s *SQL for Smarties* (see earlier reference), and to derive a broader list of "matched" values, programmers should make every attempt to develop a comprehensive list of search values to widen the scope of possibilities. We can expand our original search criteria in the previous example to include the missing possibilities using OR logic.

SQL Code

```
PROC SQL;
  SELECT CUSTNUM,
         CUSTNAME,
         CUSTCITY
    FROM CUSTOMERS2
      WHERE CUSTNAME =* 'Lafler'   OR
            CUSTNAME =* 'Laughler' OR
            CUSTNAME =* 'Lasler';
QUIT;
```

Results

```
                    The SAS System

        CUSTNUM  CUSTNAME          CUSTCITY
        ______________________________________
        7        Lafler            Spring Valley
        7        Loffler           Spring Valley
        7        Laughler          Spring Valley
        7        Laffler           Spring Valley
```

2.3.4.10 Finding the First Nonmissing Value

The first example provides a way to find the first nonmissing value in a column or list. Specified in a SELECT statement, the **COALESCE** function inspects a column, or, in the case of a list, scans the arguments from left to right, and returns the first nonmissing or non-null value. If all values are missing, the result is missing. To take advantage of the COALESCE function, use arguments all of the same data type. The next example illustrates one approach to computing the total cost for each product purchased based on the number of units and unit costs columns in the PURCHASES table. If either the UNITS or UNITCOST columns contain a missing value, a zero is assigned to prevent the propagation of missing values.

SQL Code

```
PROC SQL;
  SELECT CUSTNUM,
         ITEM,
         UNITS,
         UNITCOST,
         (COALESCE(UNITS, 0) * COALESCE(UNITCOST, 0))
               AS Totcost FORMAT=DOLLAR6.2
    FROM PURCHASES;
QUIT;
```

Results

The SAS System

Custnum	Item	Units	Unitcost	Totcost
1	Chair	1	$179.00	179.00
1	Pens	12	$0.89	$10.68
1	Paper	4	$6.95	$27.80
1	Stapler	1	$8.95	$8.95
7	Mouse Pad	1	$11.79	$11.79
7	Pens	24	$1.59	$38.16
13	Markers	.	$0.99	$0.00

2.3.4.11 Producing a Row Number

A unique undocumented, but unsupported, feature for producing a row (observation) count can be obtained with the **MONOTONIC( )** function. Similar to the row numbers produced and displayed on output from the PRINT procedure (without the NOOBS option specified), the MONOTONIC() function displays row numbers too. The MONOTONIC() function automatically creates a column (variable) in the output results or in a new table. Because this is an undocumented feature and not supported in the SQL procedure, you are cautioned that it is possible to obtain duplicates or missing values with the MONOTONIC() function. The next example illustrates the creation of a row number using the MONOTONIC() function in a SELECT statement.

SQL Code

```
PROC SQL;
  SELECT MONOTONIC() AS Row_Number FORMAT=COMMA6.,
         ITEM,
         UNITS,
         UNITCOST
    FROM PURCHASES;
QUIT;
```

Results

The SAS System

Row_Number	Item	Units	Unitcost
1	Chair	1	$179.00
2	Pens	12	$0.89
3	Paper	4	$6.95
4	Stapler	1	$8.95
5	Mouse Pad	1	$11.79
6	Pens	24	$1.59
7	Markers	.	$0.99

A row number can also be produced with the documented and supported SQL procedure option **NUMBER**. Unlike the MONOTONIC() function, the NUMBER option does not create a new column in a new table. The NUMBER option is illustrated below.

SQL Code

```
PROC SQL NUMBER;
  SELECT ITEM,
         UNITS,
         UNITCOST
    FROM PURCHASES;
QUIT;
```

Results

```
                    The SAS System

        Row  Item               Units        Unitcost
        ______________________________________________

          1  Chair                  1         $179.00
          2  Pens                  12           $0.89
          3  Paper                  4           $6.95
          4  Stapler                1           $8.95
          5  Mouse Pad              1          $11.79
          6  Pens                  24           $1.59
          7  Markers                .           $0.99
```

2.3.5 Summarizing Data

The SQL procedure is a wonderful tool for summarizing (or aggregating) data. It provides a number of useful summary (or aggregate) functions to help perform calculations, descriptive statistics, and other aggregating operations in a SELECT statement or HAVING clause. These functions are designed to summarize information and not display detail about data.

Without the availability of summary functions, you would have to construct the necessary logic using somewhat complicated SQL programming constructs. When using a summary function without a GROUP BY clause (see Chapter 3 for details), all the rows in a table are treated as a single group. Consequently, the results are often a single row value.

A number of summary functions are available including facilities to count nonmissing values; determine the minimum and maximum values in specific columns; return the range of values; compute the mean, standard deviation, and variance of specific values; and perform other aggregating functions. The following table is an alphabetical list of the available summary functions. When multiple names for the same function are available, the ANSI-approved name appears first.

Summary Functions

Summary Function	Description
AVG, MEAN	Average or mean of values
COUNT, FREQ, N	Aggregate number of nonmissing values
CSS	Corrected sum of squares
CV	Coefficient of variation
MAX	Largest value
MIN	Smallest value
NMISS	Number of missing values
PRT	Probability of a greater absolute value of Student's *t*
RANGE	Difference between the largest and smallest values
STD	Standard deviation
STDERR	Standard error of the mean
SUM	Sum of values
SUMWGT	Sum of the weight variable values, which is 1
T	Testing the hypothesis that the population mean is zero
USS	Uncorrected sum of squares
VAR	Variance

The next example uses the **COUNT** function with the **(*)** argument to produce a total number of rows, regardless if data is missing. The asterisk (*) is specified as the argument to the COUNT function to count all rows in the PURCHASES table.

SQL Code

```
PROC SQL;
  SELECT COUNT(*) AS Row_Count
    FROM PURCHASES;
QUIT;
```

Results

```
                The SAS System

                  Row_Count
                  ---------
                          7
```

Unlike the COUNT(*) function syntax that counts all rows, regardless if data is missing or not, the next example uses the **COUNT** function with the (column-name) argument to produce a total number of nonmissing rows based on the column UNITS.

SQL Code

```
PROC SQL;
  SELECT COUNT(UNITS) AS Non_Missing_Row_Count
    FROM PURCHASES;
QUIT;
```

Results

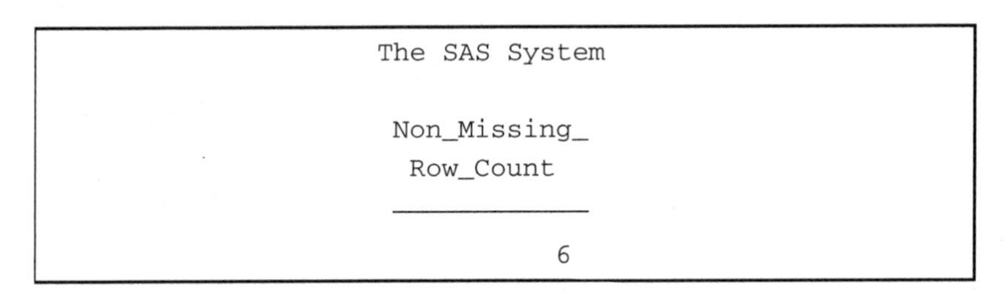

```
                The SAS System

                 Non_Missing_
                   Row_Count
                 ------------
                            6
```

The **MIN** summary function can be specified to determine what the least expensive product is in the PRODUCTS table.

SQL Code

```
PROC SQL;
  SELECT MIN(prodcost) AS Cheapest
            Format=dollar9.2 Label='Least Expensive'
    FROM PRODUCTS;
QUIT;
```

Results

```
                    The SAS System

                        Least
                    Expensive
                    _________
                       $35.00
```

In the next example, the SUM function is specified to sum numeric data types for a selected column. Suppose you wanted to determine the total costs of all purchases by customers who bought pens and markers. You could construct the following query to sum all nonmissing values for customers who purchased pens and markers in the PURCHASES table as follows.

SQL Code

```
PROC SQL;
  SELECT SUM((UNITS) * (UNITCOST))
              AS Total_Purchases FORMAT=DOLLAR6.2
    FROM PURCHASES
      WHERE UPCASE(ITEM)='PENS' OR
            UPCASE(ITEM)='MARKERS';
QUIT;
```

Results

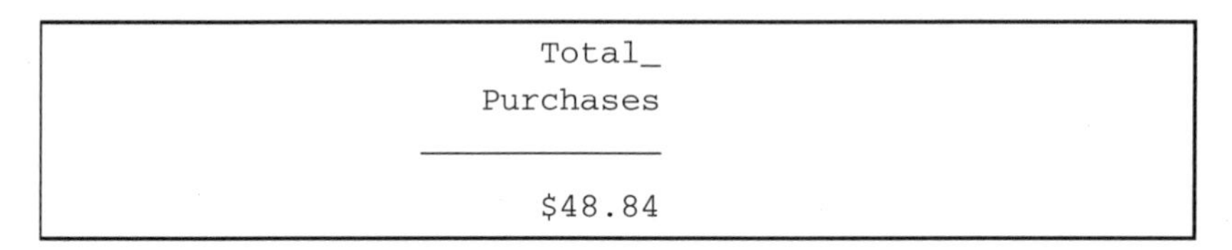

```
Total_
Purchases
_________
$48.84
```

Data can also be summarized down rows (observations) as well as across columns (variables). This flexibility gives SAS users an incredible range of power and the ability to take advantage of several SAS-supplied (or built-in) summary functions. These techniques permit the average of quantities rather than the set of all quantities. Without the ability to summarize data in PROC SQL, users would be forced to write complicated formulas and/or routines or even write and test DATA step programs to summarize data. Two examples illustrate how SQL can be constructed to summarize data: 1) summarizing data down rows and 2) summarizing data across columns.

1. Summarizing data down rows

The SQL procedure can be used to produce a single aggregate value by summarizing data down rows (or observations). The advantage of using a summary function in PROC SQL is that it will generally compute the aggregate quicker than if a user-defined equation were constructed. It also saves the effort of having to construct and test a program containing the user-defined equation in the first place. Suppose you wanted to know the average product cost for all software in the PRODUCTS table containing a variety of products. The following query computes the average product cost and produces a single aggregate value using the AVG function.

SQL Code

```
PROC SQL;
 SELECT AVG(PRODCOST) AS
      Average_Product_Cost
  FROM PRODUCTS
   WHERE UPCASE(PRODTYPE) IN
         ('SOFTWARE');
QUIT;
```

Results

```
              Average_
          Product Cost
                   324
```

2. Summarizing data across columns

When a computation is needed on two or more columns in a row, the SQL procedure can be used to summarize data across columns. Suppose you wanted to know the average cost of products in inventory. The next example computes the average inventory cost for each product without using a summary function, and once computed displays the value for each row as Average_Price.

SQL Code

```
PROC SQL;
 SELECT PRODNUM,
        (INVPRICE / INVQTY) AS
          Averge_Price
            FORMAT=DOLLAR8.2
  FROM INVOICE;
QUIT;
```

Results

```
      Product
       Number  Averge_Price
  ______________________

         5001       $299.00
         6001       $799.00
         2101        $35.00
         1110      $3200.00
         5002       $399.00
         6000        $99.00
         1200      $3300.00
```

2.3.6 Predicates

Predicates are used in PROC SQL to perform direct comparisons between two conditions or expressions. Six predicates will be looked at:

- BETWEEN
- IN
- IS NULL, IS MISSING
- LIKE
- EXISTS

2.3.6.1 Selecting a Range of Values

The **BETWEEN** predicate is a way of simplifying a query by selecting column values within a designated range of values. BETWEEN is equivalent to one LE (less than or equal) and one GE (greater than or equal) condition being ANDed together. It is extremely flexible because it works with character, numeric, and date values. Programmers can also combine two or more BETWEEN predicates with AND or OR operators for more complicated conditions. In the next example a range of products costing between $200 and $500 inclusively are selected from the PRODUCTS table.

SQL Code

```
PROC SQL;
  SELECT PRODNAME,
         PRODTYPE,
         PRODCOST
    FROM PRODUCTS
      WHERE PRODCOST BETWEEN 200 AND 500;
QUIT;
```

Results

```
                    The SAS System

                                                   Product
Product Name                Product Type             Cost
------------------------------------------------------------
Spreadsheet Software        Software              $299.00
Database Software           Software              $399.00
Wordprocessor Software      Software              $299.00
Graphics Software           Software              $299.00
```

In the next example, products are selected from the INVENTORY table that were ordered between the years 1999 and 2000. The YEAR function returns the year portion from a SAS date value and is used as the range of values in the WHERE clause.

SQL Code

```
PROC SQL;
  SELECT PRODNUM,
         INVENQTY,
         ORDDATE
    FROM INVENTORY
      WHERE YEAR(ORDDATE) BETWEEN 1999 AND 2000;
QUIT;
```

Results

The SAS System

Product Number	Inventory Quantity	Date Inventory Last Ordered
1110	20	09/01/2000
1700	10	08/15/2000
5001	5	08/15/2000
5002	3	08/15/2000
5003	10	08/15/2000
5004	20	09/01/2000
5001	2	09/01/2000

The BETWEEN predicate and OR operator are used together in the next example to select products ordered between 1999 and 2000 or where inventory quantities are greater than 15. The YEAR function returns the year portion from a SAS date value and is used as the range of values in the WHERE clause.

SQL Code

```
PROC SQL;
  SELECT PRODNUM,
         INVENQTY,
         ORDDATE
    FROM INVENTORY
      WHERE (YEAR(ORDDATE) BETWEEN 1999 AND 2000) OR
               INVENQTY > 15;
QUIT;
```

Results

```
                    The SAS System

                                           Date
                                      Inventory
             Product   Inventory           Last
              Number    Quantity        Ordered
           ------------------------------------
                1110          20     09/01/2000
                1700          10     08/15/2000
                5001           5     08/15/2000
                5002           3     08/15/2000
                5003          10     08/15/2000
                5004          20     09/01/2000
                5001           2     09/01/2000
```

2.3.6.2 Selecting Nonconsecutive Values

The **IN** predicate selects one or more rows based on the matching of one or more column values in a set of values. The IN predicate creates an OR condition between each value and returns a Boolean value of true if a column value is equal to one or more of the values in the expression list. Although the IN predicate can be specified with single column values, it may be less costly to specify the "=" sign instead. In the next example, the "=" sign is used rather than the IN predicate to select phones from the PRODUCTS table.

SQL Code

```
PROC SQL;
  SELECT PRODNAME,
         PRODTYPE,
         PRODCOST
    FROM PRODUCTS
      WHERE UPCASE(PRODTYPE) = 'PHONE';
QUIT;
```

Results

The SAS System

Product Name	Product Type	Product Cost
Analog Cell Phone	Phone	$35.00
Digital Cell Phone	Phone	$175.00
Office Phone	Phone	$130.00

In the next example, both phones and software products are selected from the PRODUCTS table. To avoid having to specify two OR conditions, you can specify the IN predicate.

SQL Code

```
PROC SQL;
  SELECT PRODNAME,
         PRODTYPE,
         PRODCOST
    FROM PRODUCTS
      WHERE UPCASE(PRODTYPE) IN ('PHONE', 'SOFTWARE');
QUIT;
```

Results

```
                     The SAS System

                                                    Product
Product Name                  Product Type             Cost
___________________________________________________________

Analog Cell Phone             Phone                  $35.00
Digital Cell Phone            Phone                 $175.00
Office Phone                  Phone                 $130.00
Spreadsheet Software          Software              $299.00
Database Software             Software              $399.00
Wordprocessor Software        Software              $299.00
Graphics Software             Software              $299.00
```

2.3.6.3 Testing for Null or Missing Values

The **IS NULL** predicate is the ANSI approach to selecting one or more rows by evaluating whether a column value is missing or null (see earlier section on null values). The next example selects products from the INVENTORY table that are out-of-stock in inventory.

SQL Code

```
PROC SQL;
  SELECT PRODNUM,
         INVENQTY,
         INVENCST
    FROM INVENTORY
      WHERE INVENQTY IS NULL;
QUIT;
```

SAS Log Results

```
     PROC SQL;
       SELECT PRODNUM,
              INVENQTY,
              INVENCST
         FROM INVENTORY
           WHERE INVENQTY IS NULL;
NOTE: No rows were selected.
     QUIT;
NOTE: PROCEDURE SQL used:
      real time           0.05 seconds
```

In the next example products are selected from the INVENTORY table that are currently stocked in inventory. Note that the predicates NOT IS NULL or IS NOT NULL can be specified to produce the same results.

SQL Code

```
PROC SQL;
  SELECT PRODNUM,
         INVENQTY,
         INVENCST
    FROM INVENTORY
      WHERE INVENQTY IS NOT NULL;
QUIT;
```

Results

```
                  The SAS System

         Product  Inventory   Inventory
          Number   Quantity        Cost
         ______________________________
            1110         20  $45,000.00
            1700         10  $28,000.00
            5001          5   $1,000.00
            5002          3     $900.00
            5003         10   $2,000.00
            5004         20   $1,400.00
            5001          2   $1,200.00
```

The IS MISSING predicate performs identically to the IS NULL predicate by selecting one or more rows containing a missing value (null). The only difference is that specifying IS NULL is the ANSI standard way of expressing the predicate and IS MISSING is commonly used in the SAS System.

The next example uses the IS MISSING predicate with the NOT predicates to select products from the INVENTORY table that are stocked in inventory.

SQL Code

```
PROC SQL;
  SELECT PRODNUM,
         INVENQTY,
         INVENCST
    FROM INVENTORY
      WHERE INVENQTY IS NOT MISSING;
QUIT;
```

2.3.6.4 Finding Patterns in a String (Pattern Matching % and _)

Constructing specific search patterns in string expressions is a simple process with the LIKE predicate. The % acts as a wildcard character representing any number of characters, including any combination of upper- or lowercase characters. Combining the LIKE predicate with the % (percent sign) permits case-sensitive searches and is a popular technique used by savvy SQL programmers to find patterns in their data.

Using the LIKE operator with the % (percent sign) provides a wildcard capability enabling the selection of table rows that match a specific pattern. The LIKE predicate is case-sensitive and should be used with care. The wildcard character % preceding and following the search word selects all product types with "Soft" in the name. The following WHERE clause finds patterns in product name (PRODNAME) containing the uppercase character "A" in the first position followed by any number of characters.

SQL Code

```
PROC SQL;
  SELECT PRODNAME
    FROM PRODUCTS
      WHERE PRODNAME LIKE 'A%';
QUIT;
```

Results

```
                    The SAS System

             Product Name
             ________________________
             Analog Cell Phone
```

The next example selects products whose name contains the word "Soft". The resulting output contains product types such as "Software" and any other products containing the word "Soft".

SQL Code

```
PROC SQL;
  SELECT PRODNAME,
         PRODTYPE,
         PRODCOST
    FROM PRODUCTS
      WHERE PRODTYPE LIKE '%Soft%';
QUIT;
```

Results

```
                    The SAS System

                                                      Product
Product Name                 Product Type                Cost
-------------------------------------------------------------
Spreadsheet Software         Software                 $299.00
Database Software            Software                 $399.00
Wordprocessor Software       Software                 $299.00
Graphics Software            Software                 $299.00
```

In the next example, the LIKE predicate is used to check a column for the existence of trailing blanks. The wildcard character % followed by a blank space is specified as the search argument.

SQL Code

```
PROC SQL;
  SELECT PRODNAME
    FROM PRODUCTS
      WHERE PRODNAME LIKE  '% ';
QUIT;
```

Results

```
                    The SAS System

               Product Name
               ________________________
               Dream Machine
               Business Machine
               Travel Laptop
               Analog Cell Phone
               Digital Cell Phone
               Office Phone
               Spreadsheet Software
               Database Software
               Wordprocessor Software
               Graphics Software
```

When a pattern search for a specific number of characters is needed, using the LIKE predicate with the underscore (_) provides a way to pattern match character by character. Thus, a single underscore (_) in a specific position acts as a wildcard placement holder for that position only. Two consecutive underscores (__) act as a wildcard placement holder for those two positions. Three consecutive underscores act as a wildcard placement holder for those three positions. And so forth. In the next example, the first position used to search product type contains the character “P” and the next five positions (represented with five underscores) act as a placeholder for any value.

SQL Code

```
PROC SQL;
   SELECT PRODNAME,
          PRODTYPE,
          PRODCOST
     FROM PRODUCTS
       WHERE UPCASE(PRODTYPE) LIKE 'P_____';
QUIT;
```

Results

```
                         The SAS System

                                                          Product
Product Name                        Product Type             Cost
-----------------------------------------------------------------
Analog Cell Phone                   Phone                  $35.00
Digital Cell Phone                  Phone                 $175.00
Office Phone                        Phone                 $130.00
```

The next example illustrates a pattern search of product name (PRODNAME) where the first three positions are represented as a wildcard; the fourth position contains the lowercase character "a", followed by any combination of upper- or lowercase characters.

SQL Code

```
PROC SQL;
  SELECT PRODNAME
    FROM PRODUCTS
      WHERE PRODNAME LIKE  '___a%';
QUIT;
```

Results

```
                         The SAS System

               Product Name
               --------------------------
               Dream Machine
               Database Software
```

2.3.6.5 Testing for the Existence of a Value

The EXISTS predicate is used to test for the existence of a set of values. In the next example, a subquery is used to check for the existence of customers in the CUSTOMERS table with purchases from the PURCHASES table. More details on subqueries will be presented in Chapter 7, "Coding Complex Queries."

SQL Code

```
PROC SQL;
  SELECT CUSTNUM,
         CUSTNAME,
         CUSTCITY
    FROM CUSTOMERS2 C
      WHERE EXISTS
        (SELECT *
          FROM PURCHASES P
            WHERE C.CUSTNUM = P.CUSTNUM);
QUIT;
```

Results

```
                          The SAS System

   Customer
     Number  Customer Name                 Customer's Home City
   ----------------------------------------------------------------
          1  Smith                         San Diego
          7  Lafler                        Spring Valley
         13  Thompson                      Miami
          7  Loffler                       Spring Valley
          1  Smithe                        San Diego
          7  Laughler                      Spring Valley
          7  Laffler                       Spring Valley
```

2.4 Dictionary Tables

The SAS System generates and maintains valuable information at runtime about SAS libraries, data sets, catalogs, indexes, macros, system options, titles, and views in a collection of read-only tables called dictionary tables. Although called tables, dictionary tables are not real tables at all. Information is automatically generated at runtime and the tables' contents are made available once a SAS session is started.

Dictionary tables and their contents permit a SAS session's activities to be easily accessed and monitored. This becomes particularly important when building intelligent software applications. You can also specify other SELECT statement clauses, such as WHERE, GROUP BY, HAVING, and ORDER BY, when accessing dictionary tables. The available dictionary tables and their contents are described in the following table.

Dictionary Table Name	Contents
DICTIONARY.CATALOGS	SAS catalogs
DICTIONARY.COLUMNS	Data set columns and attributes
DICTIONARY.EXTFILES	Allocated filerefs and external physical paths
DICTIONARY.INDEXES	Data set indexes
DICTIONARY.MACROS	Global and automatic macro variables
DICTIONARY.MEMBERS	SAS data sets and other member types
DICTIONARY.OPTIONS	Current SAS System option settings
DICTIONARY.TABLES	SAS data sets and views
DICTIONARY.TITLES	Title and footnote definitions
DICTIONARY.VIEWS	SAS data views

2.4.1 Displaying Dictionary Table Definitions

You can view a dictionary table's definition and enhance your understanding of each table's contents by specifying a DESCRIBE TABLE statement. The results of the statements used to create each dictionary table can be displayed on the SAS log. For example, a DESCRIBE TABLE statement is illustrated below to display the CREATE TABLE statement used in building the OPTIONS dictionary table containing current SAS System option settings.

SQL Code

```
PROC SQL;
  DESCRIBE TABLE
    DICTIONARY.OPTIONS;
QUIT;
```

SAS Log Results

```
create table DICTIONARY.OPTIONS
  (
   optname char(32) label='Option Name',
   setting char(1024) label='Option Setting',
   optdesc char(160) label='Option Description',
   level char(8) label='Option Location'
  );
```

Note: The information contained in dictionary tables is also available to DATA and PROC steps outside the SQL procedure. Referred to as dictionary views, each view is prefaced with the letter "V" and may be shortened with abbreviated names. A dictionary view can be accessed by referencing the view by its name in the SASHELP library. Refer to the *SAS Procedures Guide* for further details on accessing and using dictionary views in the SASHELP library.

2.4.2 Dictionary Table Column Names

To help you become familiar with each dictionary table's and dictionary view's column names and their definitions, the following table identifies each unique column name, type, length, format, informat, and label.

DICTIONARY.CATALOGS or SASHELP.VCATALG

Column	Type	Length	Format	Informat	Label
Libname	char	8			Library Name
Memname	char	32			Member Name
Memtype	char	8			Member Type
Objname	char	32			Object Name
Objtype	char	8			Object Type
Objdesc	char	256			Description
Created	num		DATETIME.	DATETIME.	Date Created
Modified	num		DATETIME.	DATETIME.	Date Modified
Alias	char	8			Object Alias

DICTIONARY.COLUMNS or SASHELP.VCOLUMN

Column	Type	Length	Label
Libname	char	8	Library Name
Memname	char	32	Member Name
Memtype	char	8	Member Type
Name	char	32	Column Name
Type	char	4	Column Type
Length	num		Column Length
Npos	num		Column Position
Varnum	num		Column Number in Table
Label	char	256	Column Label
Format	char	16	Column Format
Informat	char	16	Column Informat
Idxusage	char	9	Column Index Type

DICTIONARY.EXTFILES or SASHELP.VEXTFL

Column	Type	Length	Label
Fileref	char	8	Fileref
Xpath	char	1024	Path Name
Xengine	char	8	Engine Name

DICTIONARY.INDEXES or SASHELP.VINDEX

Column	Type	Length	Label
Libname	char	8	Library Name
Memname	char	32	Member Name
Memtype	char	8	Member Type
Name	char	32	Column Name
Idxusage	char	9	Column Index Type
Indxname	char	32	Index Name
Indxpos	num		Position of Column in Concatenated Key
Nomiss	char	3	Nomiss Option
Unique	char	3	Unique Option

DICTIONARY.MACROS or SASHELP.VMACRO

Column	Type	Length	Label
Scope	char	9	Macro Scope
Name	char	32	Macro Variable Name
Offset	num		Offset into Macro Variable
Value	char	200	Macro Variable Value

DICTIONARY.MEMBERS or SASHELP.VMEMBER

Column	Type	Length	Label
Libname	char	8	Library Name
Memname	char	32	Member Name
Memtype	char	8	Member Type
Engine	char	8	Engine Name
Index	char	32	Indexes
Path	char	1024	Path Name

DICTIONARY.OPTIONS or SASHELP.VOPTION

Column	Type	Length	Label
Optname	char	32	Option Name
Setting	char	1024	Option Setting
Optdesc	char	160	Option Description
Level	char	8	Option Location

DICTIONARY.TABLES or SASHELP.VTABLE

Column	Type	Length	Format	Informat	Label
Libname	char	8			Library Name
Memname	char	32			Member Name
Memtype	char	8			Member Type
Memlabel	char	256			Dataset Label
Typemem	char	8			Dataset Type
Crdate	num		DATETIME.	DATETIME.	Date Created
Modate	num		DATETIME.	DATETIME.	Date Modified
Nobs	num				# of Obs
Obslen	num				Obs Length
Nvar	num				# of Variables
Protect	char	3			Type of Password Protection
Compress	char	8			Compression Routine
Encrypt	char	8			Encryption
Npage	num				# of Pages
Pcompress	num				% Compression
Reuse	char	3			Reuse Space
Bufsize	num				Bufsize
Delobs	num				# of Deleted Obs
Indxtype	char	9			Type of Indexes
Datarep	char	32			Data Representation
Reqvector	char	24	$HEX.	$HEX.	Requirements Vector

DICTIONARY.TITLES or SASHELP.VTITLE

Column	Type	Length	Label
Type	char	1	Title Location
Number	num		Title Number
Text	char	256	Title Text

DICTIONARY.VIEWS or SASHELP.VVIEW

Column	Type	Length	Label
Libname	char	8	Library Name
Memname	char	32	Member Name
Memtype	char	8	Member Type
Engine	char	8	Engine Name

2.4.3 Accessing a Dictionary Table's Contents

You access the content of a dictionary table with the SQL procedure's SELECT statement FROM clause. Results are displayed as rows and columns in a table. The results can be used in handling common data processing tasks including obtaining a list of allocated libraries, catalogs, and data sets, as well as communicating SAS environment settings to custom software applications. You'll want to take the time to explore the capabilities of these read-only dictionary tables and become familiar with the type of information they provide.

2.4.3.1 Dictionary.CATALOGS

Obtaining detailed information about catalogs and their members is quick and easy with the CATALOGS dictionary table. You can capture an ordered list of catalog information by member name including object name and type, description, date created and last modified, and object alias from any SAS library. For example, the following code produces a listing of the catalog objects in the SASUSER library.

Note: Because this dictionary table produces a considerable amount of information, always specify a WHERE clause when using.

SQL Code

```
PROC SQL;
  SELECT *
    FROM DICTIONARY.CATALOGS
      WHERE LIBNAME=`SASUSER';
QUIT;
```

Results

Library Name	Member Name	Member Type	Object Name	Object Type	Object Description	Date Created	Date Modified	Object Alias	Library Concatenation Level
SASUSER	PROFILE	CATALOG	VT_PRINTLIST	SLIST	vt_printlist.SLIST	21JUL04:18:26:46	21JUL04:18:26:46		0
SASUSER	PROFILE	CATALOG	FL1	SOURCE		12MAY04:00:36:30	12MAY04:00:36:30		0
SASUSER	PROFILE	CATALOG	MRUWSAVE	WSAVE		12JUL04:20:17:40	12JUL04:20:17:40		0
SASUSER	PROFILE	CATALOG	V9TUTDLG	WSAVE		21JUL04:17:42:55	21JUL04:17:42:55		0
SASUSER	PROFILE	CATALOG	VIEWWSAVE	WSAVE	Preferences View save information.	02FEB04:14:12:12	02FEB04:14:12:12		0
SASUSER	PROFILE2	CATALOG	WNTPRINT	WINPRINT		21JUL04:17:42:27	21JUL04:17:42:27		0

2.4.3.2 Dictionary.COLUMNS

Retrieving information about the columns in one or more data sets is easy with the COLUMNS dictionary table. You can capture column-level information including column name, type, length, position, label, format, informat, and indexes, as well as produce cross-reference listings containing the location of columns in a SAS library. For example, the following code requests a cross-reference listing of the tables containing the CUSTNUM column in the WORK library.

Note: Use care when specifying multiple functions with the WHERE clause because the SQL Optimizer is unable to optimize the query resulting in all allocated SAS session librefs being searched. This can cause the query to run much longer than expected.

SQL Code

```
PROC SQL;
  SELECT *
    FROM DICTIONARY.COLUMNS
      WHERE UPCASE(LIBNAME)='WORK' AND
            UPCASE(NAME)= 'CUSTNUM';
QUIT;
```

Results

Library Name	Member Name	Member Type	Column Name	Column Type	Column Length	Column Position	Column Number in Table	Column Label
WORK	CUSTOMERS	DATA	custnum	num	3	0	1	Customer Number
WORK	INVOICE	DATA	custnum	num	3	6	3	Customer Number
WORK	PURCHASES	DATA	custnum	num	4	0	1	

Column Format	Column Informat	Column Index Type	Order in Key Sequence	Extended Type	Not NULL?	Precision	Scale	Transcoded?
			0	num	no	.	.	yes
			0	num	no	.	.	yes
			0	num	no	.	.	yes

2.4.3.3 Dictionary.EXTFILES

Accessing allocated external files by fileref and corresponding physical path name information is a snap with the EXTFILES dictionary table. The results from this handy table can be used in an application to communicate whether a specific fileref has been allocated with a FILENAME statement. For example, the following code produces a listing of each individual path name by fileref.

SQL Code

```
PROC SQL;
  SELECT *
    FROM DICTIONARY.EXTFILES;
QUIT;
```

Results

Fileref	Path Name	Engine Name
#LN00003	TERMINAL	
#LN00005	C:\Documents and Settings\Valued Sony Customer\My Documents	
#LN00007	C:\Documents and Settings\Valued Sony Customer\Desktop	
#LN00014	d:\books\Proc sql - beyond the basics using sas (sas institute)\sas output\Chpt 2 - EXTFILES Dictionary Table.html	
#LN00001	D:\SAS Version 9.1\core\sasmsg	
#LN00002	TERMINAL	

2.4.3.4 Dictionary.INDEXES

It is sometimes useful to display the names of existing simple and composite indexes, or their SAS tables, that reference a specific column name. The INDEXES dictionary table provides important information to help identify indexes that improve a query's performance. For example, to display indexes that reference the CUSTNUM column name in any of the example tables, specify the following code.

Note: Readers are referred to Chapter 10, "Tuning for Performance and Efficiency," for performance tuning techniques as they relate to indexes.

SQL Code

```
PROC SQL;
  SELECT *
    FROM DICTIONARY.INDEXES
      WHERE UPCASE(NAME)='CUSTNUM'       /* Column Name  */
            AND UPCASE(LIBNAME)='WORK'; /* Library Name */
QUIT;
```

2.4.3.5 Dictionary.MACROS

The ability to capture macro variable names and their values is available with the MACROS dictionary table. The MACROS dictionary table provides information for global and macro variables, but not local macro variables. For example, to obtain columns specific to macros such as global macros SQLOBS, SQLOOPS, SQLXOBS, or SQLRC, specify the following code.

SQL Code

```
PROC SQL;
  SELECT *
    FROM DICTIONARY.MACROS
      WHERE UPCASE(SCOPE)='GLOBAL';
QUIT;
```

Results

Macro Scope	Macro Variable Name	Offset into Macro Variable	Macro Variable Value
GLOBAL	SQLOBS	0	0
GLOBAL	SQLOOPS	0	0
GLOBAL	SQLXOBS	0	0
GLOBAL	SQLRC	0	0

2.4.3.6 Dictionary.MEMBERS

Accessing a detailed list of data sets, views, and catalogs is the hallmark of the MEMBERS dictionary table. You will be able to access a terrific resource of information by library, member name and type, engine, indexes, and physical path name. For example, to obtain a list of the individual files in the WORK library, the following code is specified.

SQL Code

```
PROC SQL;
  SELECT *
    FROM DICTIONARY.MEMBERS
      WHERE UPCASE(LIBNAME)='WORK';
QUIT;
```

Results

Library Name	Member Name	Member Type	DBMS Member Type	Engine Name	Indexes	Path Name
WORK	CUSTOMERS	DATA		V9	no	D:\SAS Version 9.1\SAS Temporary Files_TD1816
WORK	INVENTORY	DATA		V9	no	D:\SAS Version 9.1\SAS Temporary Files_TD1816
WORK	INVOICE	DATA		V9	no	D:\SAS Version 9.1\SAS Temporary Files_TD1816
WORK	MANUFACTURERS	DATA		V9	no	D:\SAS Version 9.1\SAS Temporary Files_TD1816
WORK	PRODUCTS	DATA		V9	no	D:\SAS Version 9.1\SAS Temporary Files_TD1816
WORK	PURCHASES	DATA		V9	no	D:\SAS Version 9.1\SAS Temporary Files_TD1816

2.4.3.7 Dictionary.OPTIONS

The OPTIONS dictionary table provides a list of the current SAS System session's option settings including the option name, its setting, description, and location. Obtaining option settings is as easy as 1-2-3. Simply submit the following SQL query referencing the OPTIONS dictionary table as follows. A partial listing of the results from the OPTIONS dictionary table is displayed below in rich text format.

SQL Code

```
PROC SQL;
  SELECT *
    FROM DICTIONARY.OPTIONS;
QUIT;
```

Results

Option Name	Option type	Option Setting	Option Description	Option Location	Option Group
APPLETLOC	char	D:\applets\9.1	Location of Java applets	Portable	ENVFILES
ARMAGENT	char		ARM Agent to use to collect ARM records	Portable	PERFORMANCE
ARMLOC	char	ARMLOC.LOG	Identify location where ARM records are to be written	Portable	PERFORMANCE
ARMSUBSYS	char	(ARM_NONE)	Enable/Disable ARMing of SAS subsystems	Portable	PERFORMANCE
ASYNCHIO	boolean	NOASYNCHIO	Enable asynchronous input/output	Portable	SASFILES
AUTOSAVELOC	char		Identifies the location where program editor contents are auto saved	Portable	ENVDISPLAY
AUTOSIGNON	boolean	NOAUTOSIGNON	SAS/CONNECT remote submit will automatically attempt to SIGNON	Portable	COMMUNICATIONS
BATCH	boolean	NOBATCH	Use the batch set of default values for SAS system options	Portable	EXECMODES
BINDING	char	DEFAULT	Controls the binding edge for duplexed output	Portable	ODSPRINT
BOTTOMMARGIN	char	0.000 IN	Bottom margin for printed output	Portable	ODSPRINT
BUFNO	num	1	Number of buffers for each SAS data set	Portable	PERFORMANCE
BUFSIZE	num	0	Size of buffer for page of SAS data set	Portable	PERFORMANCE

2.4.3.8 Dictionary.TABLES

When you need more information about SAS files than what the MEMBERS dictionary table provides, consider using the TABLES dictionary table. The TABLES dictionary table provides such file details as library name, member name and type, date created and last modified, number of observations, observation length, number of variables, password protection, compression, encryption, number of pages, reuse space, buffer size, number of deleted observations, type of indexes, and requirements vector. For example, to obtain a detailed list of files in the WORK library, the following code is specified.

Note: Because this dictionary table produces a considerable amount of information, users should specify a WHERE clause when using it.

SQL Code

```
PROC SQL;
  SELECT *
    FROM DICTIONARY.TABLES
      WHERE UPCASE(LIBNAME)='WORK';
QUIT;
```

Results

Library Name	Member Name	Member Type	Dataset Label	Dataset Type	Date Created	Date Modified	Number of Observations	Observation Length	Number of Variables	Type of Password Protection	Compression Routine
WORK	CUSTOMERS	DATA			27JAN00:12:26:35	01NOV99:23:15:42	28	48	3	---	NO
WORK	INVENTORY	DATA			01NOV99:22:56:05	02NOV99:00:43:58	8	20	5	---	NO
WORK	INVOICE	DATA			01NOV99:23:30:35	01NOV99:23:30:35	7	20	6	---	NO
WORK	MANUFACTURERS	DATA			27JAN00:12:26:33	06NOV99:12:53:15	6	50	4	---	NO
WORK	PRODUCTS	DATA			30NOV99:13:52:17	30NOV99:13:52:17	12	51	5	---	NO
WORK	PURCHASES	DATA			07NOV99:16:07:14	11AUG02:02:18:26	8	24	4	---	NO

Encryption	Number of Pages	Percent Compression	Reuse Space	Bufsize	Number of Deleted Observations	Type of Indexes	Data Representation	Requirements Vector
NO	1	0	no	4096	0		NATIVE	18
NO	2	0	no	4096	1		NATIVE	18
NO	1	0	no	4096	0		NATIVE	18
NO	1	0	no	4096	0		NATIVE	18
NO	1	0	no	8192	0		NATIVE	18
NO	1	0	no	4096	1		NATIVE	18

2.4.3.9 Dictionary.TITLES

The TITLES dictionary table provides a listing of the currently defined titles and footnotes in a session. The table output distinguishes between titles and footnotes using a "T" or "F" in the TITLE LOCATION column. For example, the following code displays a single title and two footnotes.

SQL Code

```
PROC SQL;
  SELECT *
    FROM DICTIONARY.TITLES;
QUIT;
```

Results

Title Location	Title Number	Title Text
T	1	Software Intelligence Corporation
F	1	Prepared July 2004

2.4.3.10 Dictionary.VIEWS

The VIEWS dictionary table provides a listing of views for selected SAS libraries. The VIEWS dictionary table displays the library name, member names and type, and engine used. For example, the following code displays a single view called VIEW_CUSTOMERS from the WORK library.

SQL Code

```
PROC SQL;
  SELECT *
    FROM DICTIONARY.VIEWS
      WHERE UPCASE(LIBNAME)='WORK';
QUIT;
```

Results

Library Name	Member Name	Member Type	Engine Name
WORK	VIEW_CUSTOMERS	VIEW	SASESQL

2.5 Summary

1. The available data types in SQL are 1) numeric and 2) character (see section 2.2).

2. When a table is created with PROC SQL, numeric columns are assigned a default length of 8 bytes (see section 2.2.1).

3. SAS tables store date and time information in the form of a numeric data type (see section 2.2.2).

4. A CHAR column stores a default of 8 characters (see section 2.2.3).

5. Missing or unknown information is supported by PROC SQL in a form known as a null value. A null value is not the same as a zero value (see section 2.2.5).

6. Comparison operators are used in the SQL procedure to compare one character or numeric value to another (see section 2.3.1).

7. Logical operators (AND, OR, and NOT) are used to connect one or more expressions together in a WHERE clause (see section 2.3.2).

8. The arithmetic operators used in the SQL procedure are the same as those used in the DATA step (see section 2.3.3).

9. Character string operators and functions are typically used with character data (see section 2.3.4).

10. Predicates are used in the SQL procedure to perform direct comparisons between two conditions or expressions (see section 2.3.6).

11. Dictionary tables provide information about the SAS environment (see section 2.4).

Chapter 3

Formatting Output

3.1 Introduction

Programmers want and expect to be able to format output in a variety of ways. With the SQL procedure programmers have forged innovative ways to enhance the appearance of output including double-spacing rows of output, concatenating two or more columns, inserting text and constants between selected columns, displaying column headers for derived fields, and much more. As a value-added feature, the SQL procedure (not part of ANSI-standard SQL — see "Preface" to this book) can be integrated with the Output Delivery System to enhance and format output in ways not otherwise available.

3.2 Formatting Output

PROC SQL consists of a standard set of statements and options to create, retrieve, alter, transform, and transfer data regardless of the operating system or where the data is located. These features provide tremendous power as well as control when integrating information from a variety of sources in a number of ways. Because emphasis is placed on PROC SQL's data manipulation capabilities and not on its format and output capabilities, many programmers are unfamiliar with the SQL procedure's output-producing features. Consequently, programmers resort to using report writers or special outputting tools to create the best looking output. To illustrate the virtues of PROC SQL in the SAS System, this chapter presents numerous examples on how output can be formatted and produced.

3.2.1 Writing a Blank Line Between Each Row

Being able to display a blank line between each row of output is available as a procedure option in PROC SQL. As with PROC PRINT, specifying **DOUBLE** in the SQL procedure inserts a blank line between each row of output (NODOUBLE is the default). Setting this option is especially useful when one or more flowed lines spans or wraps in the output because it provides visual separation between each row of data. This example illustrates using the DOUBLE option to double-space output.

SQL Code

```
PROC SQL DOUBLE;
  SELECT *
    FROM INVOICE;
QUIT;
```

Results

The SAS System

Invoice Number	Manufacturer Number	Customer Number	Invoice Quantity - Units Sold	Invoice Unit Price	Product Number
1001	500	201	5	$1,495.00	5001
1002	600	1301	2	$1,598.00	6001
1003	210	101	7	$245.00	2101
1004	111	501	3	$9,600.00	1110
1005	500	801	2	$798.00	5002
1006	500	901	4	$396.00	6000
1007	500	401	7	$23,100.00	1200

To revert back to single-spaced output, you can specify the **RESET** statement as long as the QUIT statement has not been issued to turn off the SQL procedure. When PROC SQL is active, you can specify the RESET statement with or without options to reestablish each option's original settings. When the RESET statement is specified with one or more options, only those options are reset. This example illustrates the NODOUBLE option specified to turn off double-spaced output and reset printing back to the default single-spaced output.

SQL Code

```
PROC SQL;
  RESET NODOUBLE;
QUIT;
```

3.2.2 Displaying Row Numbers

You can specify an SQL procedure option called **NUMBER** to display row numbers on output under the column heading Row. The NUMBER option displays row numbers on output. The next example shows the NUMBER option being specified to turn on the display of row numbers.

SQL Code

```
PROC SQL NUMBER;
  SELECT ITEM,
         UNITS,
         UNITCOST
    FROM PURCHASES;
QUIT;
```

Results

```
                    The SAS System

     Row  Item              Units       Unitcost
     -------------------------------------------
       1  Chair                 1        $179.00
       2  Pens                 12          $0.89
       3  Paper                 4          $6.95
       4  Stapler               1          $8.95
       5  Mouse Pad             1         $11.79
       6  Pens                 24          $1.59
       7  Markers               .          $0.99
```

3.2.3 Concatenating Character Strings

As was presented in Chapter 2, "Working with Data in PROC SQL," two or more strings can be concatenated together to produce a combined and longer string of characters. The concatenation character string operator, represented by two vertical bars "||", "!!", or "¦¦" (depending on the operating system and keyboard being used), combines two or more strings or columns together forming a new string value. The next example shows the concatenation of the manufacturer city and state columns from the MANUFACTURERS table so that the second column immediately follows the first. Although the two character strings are concatenated, the output illustrates potential problems as a result of using the concatenation operator.

First, column headers have been suppressed for both columns. Without header information, a true understanding of the contents of the output may be in jeopardy. Next, blanks are automatically padded to the entire length of the first concatenated column for each row of data resulting in a "fixed" length for the first column.

SQL Code

```
PROC SQL;
  SELECT manucity || manustat
    FROM  MANUFACTURERS;
QUIT;
```

Results

```
                The SAS System


          ____________________________
          Houston                   TX
          San Diego                 CA
          Miami                     FL
          San Mateo                 CA
          San Diego                 CA
          San Diego                 CA
```

To make the preceding output appear a bit more readable and complete, you should consider a few modifications. First, column headers could be assigned as aliases using the AS operator. The maximum size of a user-defined column header is 32 bytes (following valid SAS naming conventions). Finally, the TRIM function (described in Chapter 2,

"Working with Data in PROC SQL") could be used to remove trailing blanks in the city column. This makes the second column act as a floating field.

SQL Code

```
PROC SQL;
  SELECT TRIM(manucity) || manustat AS Headquarters
    FROM MANUFACTURERS;
QUIT;
```

Results

```
                  The SAS System

               Headquarters
               ----------------------
               HoustonTX
               San DiegoCA
               MiamiFL
               San MateoCA
               San DiegoCA
               San DiegoCA
```

Although the preceding output illustrates that some changes were made, it still is difficult to read. A few more cosmetic changes should be made to make it more aesthetically appealing and readable. In the next section, the output will be customized to give the data further separation.

3.2.4 Inserting Text and Constants Between Columns

At times, it is useful to be able to insert text and/or constants in query output. This enables special characters including symbols and comments to be inserted in the output. We can improve the output in the previous example by inserting a comma "," and a single blank space between the manufacturer city and state information. The final output illustrates an acceptable way to display columnar data using a "free-floating" presentation as opposed to fixed columns.

SQL Code

```
PROC SQL;
  SELECT trim(manucity) || ', ' || manustat
           As Headquarters
    FROM MANUFACTURERS;
QUIT;
```

Results

```
                    The SAS System

                    Headquarters
                    ---------------------
                    Houston, TX
                    San Diego, CA
                    Miami, FL
                    San Mateo, CA
                    San Diego, CA
                    San Diego, CA
```

Another method of automatically concatenating character strings, removing leading and trailing blanks, and inserting text and constants is with the **CATX** function. The next example shows how the CATX function is specified with a "," used as a separator between character strings MANUCITY and MANUSTAT.

SQL Code

```
PROC SQL;
  SELECT CATX(',', manucity, manustat)
           As Headquarters
    FROM MANUFACTURERS;
QUIT;
```

Results

```
                The SAS System

             Headquarters
             _____________________
             Houston, TX
             San Diego, CA
             Miami, FL
             San Mateo, CA
             San Diego, CA
             San Diego, CA
```

3.2.5 Using Scalar Expressions with Selected Columns

In computing terms, a scalar refers to a quantity represented by a single number or value. The value is not represented as an array or list of values, but as a single value. For example, the value 7 is a scalar value, but (0,0,7) is not. PROC SQL allows the use of scalar expressions and constants with selected columns. Typically, these expressions replace or augment one or more columns in the SELECT statement. To illustrate how a scalar expression is used, assume that a value of 7.5% representing the sales tax percentage is computed for each product in the PRODUCTS table. The results consist of the product name, product cost, and derived computed sales tax column.

Note: Although the computed column contains the results of the sales tax computation for each product, it does not contain a column heading.

SQL Code

```
PROC SQL;
  SELECT prodname, prodcost,
         .075 * prodcost
    FROM PRODUCTS;
QUIT;
```

Results

```
                    The SAS System

                                      Product
     Product Name                        Cost
     ____________________________________________________

     Dream Machine                  $3,200.00        240
     Business Machine               $3,300.00      247.5
     Travel Laptop                  $3,400.00        255
     Analog Cell Phone                 $35.00      2.625
     Digital Cell Phone               $175.00     13.125
     Office Phone                     $130.00       9.75
     Spreadsheet Software             $299.00     22.425
     Database Software                $399.00     29.925
     Wordprocessor Software           $299.00     22.425
     Graphics Software                $299.00     22.425
```

In the next two examples, a column header or alias is assigned to the derived sales tax column computed in the previous example. Two methods exist for achieving this. The first method uses the **AS** keyword to not only name the derived column, but also to permit referencing the column later in the query. This is useful in situations where a reference to the ordinal position is needed. The next example uses the ordinal position to reference a column in a query with the **ORDER BY** clause.

SQL Code

```
PROC SQL;
  SELECT prodname, prodcost,
         .075 * prodcost AS Sales_Tax
    FROM PRODUCTS
      ORDER BY 3;
QUIT;
```

Results

```
                    The SAS System

                                  Product
  Product Name                       Cost  Sales_Tax
  ________________________________________________

  Analog Cell Phone                $35.00      2.625
  Office Phone                    $130.00       9.75
  Digital Cell Phone              $175.00     13.125
  Spreadsheet Software            $299.00     22.425
  Graphics Software               $299.00     22.425
  Wordprocessor Software          $299.00     22.425
  Database Software               $399.00     29.925
  Dream Machine                 $3,200.00        240
  Business Machine              $3,300.00      247.5
  Travel Laptop                 $3,400.00        255
```

The next example illustrates the second method of assigning a column heading for the computed sales tax column using the **LABEL=** option. To further enhance the output's readability, a numeric dollar format is specified.

Note: Because the next example is only a query and the table is not being updated, the assigned attributes are only available for the duration of the step and are not permanently saved in the table's record descriptor.

SQL Code

```
PROC SQL;
  SELECT prodname, prodcost,
         .075 * prodcost FORMAT=DOLLAR7.2
                         LABEL='Sales Tax'
    FROM PRODUCTS;
QUIT;
```

Results

```
                  The SAS System

                                   Product       Sales
     Product Name                     Cost         Tax
     ____________________________________________________

     Dream Machine                $3,200.00     $240.00
     Business Machine             $3,300.00     $247.50
     Travel Laptop                $3,400.00     $255.00
     Analog Cell Phone               $35.00       $2.63
     Digital Cell Phone             $175.00      $13.13
     Office Phone                   $130.00       $9.75
     Spreadsheet Software           $299.00      $22.43
     Database Software              $399.00      $29.93
     Wordprocessor Software         $299.00      $22.43
     Graphics Software              $299.00      $22.43
```

3.2.6 Ordering Output by Columns

By definition, tables are unordered sets of data. The data that comes from a table does not automatically appear in any particular order. To offset this behavior, the SQL procedure provides the ability to impose order in a table by using an ORDER BY clause. When used, this clause orders the query results according to the values in one or more selected columns, it must be specified after the FROM clause.

Rows of data can be ordered in ascending or descending (DESC) order for each column specified (ascending is the default order). To illustrate how selected columns of data can be ordered, let's first view the PRODUCTS table and all its columns arranged in ascending order by product number (PRODNUM).

SQL Code

```
PROC SQL;
  SELECT *
    FROM PRODUCTS
      ORDER BY prodnum;
QUIT;
```

Results

The SAS System

Product Number	Product Name	Manufacturer Number	Product Type	Product Cost
1110	Dream Machine	111	Workstation	$3,200.00
1200	Business Machine	120	Workstation	$3,300.00
1700	Travel Laptop	170	Laptop	$3,400.00
2101	Analog Cell Phone	210	Phone	$35.00
2102	Digital Cell Phone	210	Phone	$175.00
2200	Office Phone	220	Phone	$130.00
5001	Spreadsheet Software	500	Software	$299.00
5002	Database Software	500	Software	$399.00
5003	Wordprocessor Software	500	Software	$299.00
5004	Graphics Software	500	Software	$299.00

The next example illustrates a query that selects and orders multiple columns of data from the PRODUCTS table. Output is arranged first in ascending order by product type (PRODTYPE) and within product type in descending order by product cost. The code and output is shown.

SQL Code

```
PROC SQL;
  SELECT prodname, prodtype, prodcost, prodnum
    FROM PRODUCTS
      ORDER BY prodtype, prodcost DESC;
QUIT;
```

Results

```
                          The SAS System

                                                Product     Product
Product Name             Product Type              Cost      Number
-------------------------------------------------------------------
Travel Laptop            Laptop               $3,400.00        1700
Digital Cell Phone       Phone                  $175.00        2102
Office Phone             Phone                  $130.00        2200
Analog Cell Phone        Phone                   $35.00        2101
Database Software        Software               $399.00        5002
Spreadsheet Software     Software               $299.00        5001
Graphics Software        Software               $299.00        5004
Wordprocessor Software   Software               $299.00        5003
Business Machine         Workstation          $3,300.00        1200
Dream Machine            Workstation          $3,200.00        1110
```

3.2.7 Grouping Data with Summary Functions

Occasionally it may be useful to display data in designated groups. A GROUP BY clause is used in these situations to aggregate and order groups of data using a designated column with the same value. When a GROUP BY clause is used without a summary function, SAS issues a warning on the SAS log with the message, "A GROUP BY clause has been transformed into an ORDER BY clause because neither the SELECT clause nor the optional HAVING clause of the associated table-expression referenced a summary function." The GROUP BY is transformed into an ORDER BY clause and then processed.

When a GROUP BY clause is used without specifying a summary function in the SELECT statement, the entire table is treated as a single group and ordered in ascending order. The next example illustrates a GROUP BY clause without any summary function specifications. Due to the absence of any summary functions, the GROUP BY clause is automatically transformed into an ORDER BY clause, with the rows being ordered in ascending order by product type (PRODTYPE).

SQL Code

```
PROC SQL;
  SELECT prodtype,
         prodcost
    FROM PRODUCTS
      GROUP BY prodtype;
QUIT;
```

Results

```
                The SAS System

                               Product
          Product Type            Cost
          ____________________________

          Laptop             $3,400.00
          Phone                $130.00
          Phone                $175.00
          Phone                 $35.00
          Software             $299.00
          Software             $299.00
          Software             $299.00
          Software             $399.00
          Workstation        $3,200.00
          Workstation        $3,300.00
```

When a GROUP BY clause is used with a summary function, the rows are aggregated in a series of groups. This means that an aggregate function is evaluated on a group of rows and not on a single row at a time. Suppose the least expensive product in each product category needs to be identified. A separate query for each product category could be specified using the MIN function to determine the cheapest product. But this would require separate runs to be executed — not a very good approach. A better way to do this would be to specify a GROUP BY clause in a single statement as follows.

SQL Code

```
PROC SQL;
  SELECT prodtype,
         MIN(prodcost) AS Cheapest
            Format=dollar9.2 Label='Least Expensive'
    FROM PRODUCTS
      GROUP BY prodtype;
QUIT;
```

Results

```
                  The SAS System

                                     Least
              Product Type       Expensive
              ----------------------------
              Laptop             $3,400.00
              Phone                 $35.00
              Software             $299.00
              Workstation        $3,200.00
```

3.2.8 Grouping Data and Sorting

In the absence of an ORDER BY clause, the SQL procedure automatically sorts the results from a grouped query in the same order as specified in the GROUP BY clause. When both an ORDER BY and GROUP BY clause are specified for the same column(s) or column order, no additional processing occurs to satisfy the request. Because the ORDER BY and GROUP BY clauses are not mutually exclusive, they can be used together. Internally, the GROUP BY clause first sorts the results on the grouping column(s) and then aggregates the rows of the query by the same grouping column.

But what happens when the column(s) specified in the ORDER BY and GROUP BY clauses are not the same? In these situations additional processing requirements are generally needed. The additional processing, in the worst case scenario, may require remerging summary statistics back with the original data. In other cases, additional sorting requirements may be necessary. Suppose information about the least expensive product in each product category is desired. But instead of automatically sorting the results in ascending order by product type, as before, the results will be displayed in ascending order by the least expensive product.

SQL Code

```
PROC SQL;
  SELECT prodtype,
         MIN(prodcost) AS Cheapest
            Format=dollar9.2 Label='Least Expensive'
    FROM PRODUCTS
      GROUP BY prodtype
        ORDER BY cheapest;
QUIT;
```

Results

```
                    The SAS System

                                        Least
              Product Type          Expensive
              ___________________________
              Phone                    $35.00
              Software                $299.00
              Workstation           $3,200.00
              Laptop                $3,400.00
```

3.2.9 Subsetting Groups with the HAVING Clause

When processing groups of data, it is frequently useful to subset aggregated rows (or groups) of data. This way, aggregated data can be filtered one group at a time in contrast to the WHERE clause where individual rows of data are filtered one row at a time, not aggregated rows. SQL provides a convenient way to subset (or filter) groups of data by using the GROUP BY and HAVING clauses. The HAVING clause is applied after the summary of all observations.

Suppose you want to identify only those product groupings that have an average cost less than $200.00 from the PRODUCTS table. Your first inclination might be to use a summary function in a WHERE clause. But this would not be valid because a WHERE clause is designed specifically to evaluate a single row at a time. This is in direct contrast with the way a summary function processes data because summary functions evaluate groups of rows at a time, not a single row of data at a time as with a WHERE clause. For those already familiar with subqueries as discussed in Chapter 7, "Coding Complex Queries," you could also approach the problem as a complex query. But there is an easier and more straightforward way of identifying and selecting the desired product groups using the **GROUP BY** and **HAVING** clauses, as follows.

SQL Code

```
PROC SQL;
  SELECT prodtype,
         AVG(prodcost)
            FORMAT=DOLLAR9.2 LABEL='Average Product Cost'
    FROM PRODUCTS
      GROUP BY prodtype
        HAVING AVG(prodcost) <= 200.00;
QUIT;
```

Results

```
                    The SAS System

                                   Average
                                   Product
               Product Type           Cost
               ---------------------------
               Phone               $113.33
```

3.3 Formatting Output with the Output Delivery System

The SAS System provides users with a familiar and automatic way to look at output in a listing file. Although easy to use, it is not extremely flexible when it comes to creating "nice" looking output. The SAS Output Delivery System (ODS) provides many ways to format output by controlling the way it is accessed and formatted. Many output formats are available with ODS, including traditional SAS monospace font (that is, listing).

ODS was first introduced in Version 7 as a way to improve the appearance of traditional SAS output. It enables "quality" looking output to be produced without the need to import it into word processors such as MS-Word and WordPerfect. Since then, many new output formatting features and options have been made available for SAS users. With

ODS, users have a powerful and easy way to create and access formatted procedure and DATA step output.

3.3.1 ODS and Output Formats

ODS statements are classified as global statements and are processed immediately by the SAS System. With built-in format engines, referred to as output destinations, ODS prepares output using special formats and layouts. The diagram below illustrates the types of output that can be produced with ODS.

ODS Output Destinations

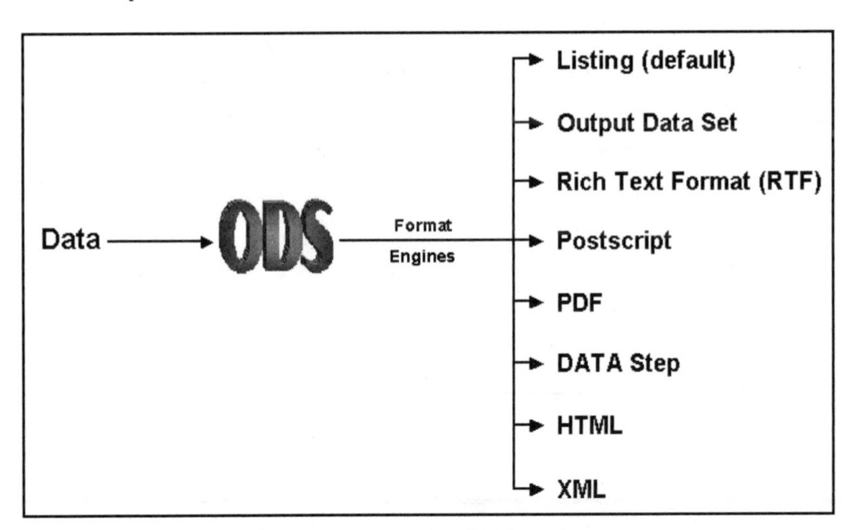

PROC and DATA steps produce output in the form of an output object to any and all open destinations. An output destination controls what format engine is turned on during a step or until you specify another ODS statement. One or more output destinations can be opened concurrently. When a destination is open, one or more output objects can be sent to it. Conversely, when closed, output objects are not sent to the destination.

Several good ODS books are available for further study on this exciting facility including *The Complete Guide to the SAS Output Delivery System, Version 8; Output Delivery System: The Basics* by Lauren E. Haworth; and *Quick Results with the Output Delivery System* by Sunil Kumar Gupta.

3.3.2 Sending Output to a SAS Data Set

Output produced by the SQL procedure can also be used as input to another vendor's SQL, procedure, or DATA step. ODS provides an easy and consistent alternative for creating a SAS table of results. For users already familiar with ODS, this approach will consist of specifying the OUTPUT destination in an ODS statement. For users preferring a more traditional ANSI SQL approach, the CREATE TABLE statement (see Chapter 5, "Creating, Populating, and Deleting Tables," for more details) will be the method of choice. An output SAS data set produced by the SQL procedure can also be used as input in another PROC SQL step, SAS procedure, or DATA step.

Although the CREATE TABLE statement is the standard method of creating a table in PROC SQL, you can also use the ODS OUTPUT statement to produce a table. The result table is then a rectangular structure consisting of one or more rows and columns. In this example, the SQL procedure's results are stored in object SQL_Results and are then sent to data set SQL_DATA using the ODS OUTPUT destination. The resulting data set is displayed using VIEWTABLE.

SQL Code

```
ODS LISTING CLOSE;
ODS OUTPUT SQL_Results = SQL_DATA;
PROC SQL;
  TITLE1 'Delivering Output to a Data Set';
  SELECT prodname, prodtype, prodcost, prodnum
    FROM PRODUCTS
      ORDER BY prodtype;
QUIT;
ODS OUTPUT CLOSE;
ODS LISTING;
```

Results

	Product Name	Product Type	Product Cost	Product Number
1	Travel Laptop	Laptop	$3,400.00	1700
2	Office Phone	Phone	$130.00	2200
3	Digital Cell Phone	Phone	$175.00	2102
4	Analog Cell Phone	Phone	$35.00	2101
5	Spreadsheet Software	Software	$299.00	5001
6	Graphics Software	Software	$299.00	5004
7	Wordprocessor Software	Software	$299.00	5003
8	Database Software	Software	$399.00	5002
9	Dream Machine	Workstation	$3,200.00	1110
10	Business Machine	Workstation	$3,300.00	1200

3.3.3 Converting Output to Rich Text Format

Rich Text Format (RTF) is text that includes codes that represent special formatting attributes. Although most frequently associated with a word processing program's ability to read and create encapsulated text fonts and highlighting attributes during copy-and-paste operations, the ODS RTF destination permits output generated by SAS to be packaged as rich text format. This enables you to produce output that can be shared.

The next example illustrates SQL output being sent to an external RTF file using the RTF format engine. First, the default Listing destination is closed, and then the RTF format engine is opened with an external file destination to which SQL results will be routed. Once the SQL procedure executes, the RTF destination is closed and the default Listing destination is opened.

Note: Opening the RTF file automatically invokes your system's default word processor and displays the file contents.

SQL Code

```
ODS LISTING CLOSE;
ODS RTF FILE='c:\SQL_Results.rtf';
PROC SQL;
  TITLE1 'Delivering Output to Rich Text Format';
  SELECT prodname, prodtype, prodcost, prodnum
    FROM PRODUCTS
      ORDER BY prodtype;
QUIT;
ODS RTF CLOSE;
ODS LISTING;
```

Results

Delivering Output to Rich Text Format

Product Name	Product Type	Product Cost	Product Number
Travel Laptop	Laptop	$3,400.00	1700
Office Phone	Phone	$130.00	2200
Digital Cell Phone	Phone	$175.00	2102
Analog Cell Phone	Phone	$35.00	2101
Spreadsheet Software	Software	$299.00	5001
Graphics Software	Software	$299.00	5004
Wordprocessor Software	Software	$299.00	5003
Database Software	Software	$399.00	5002
Dream Machine	Workstation	$3,200.00	1110
Business Machine	Workstation	$3,300.00	1200

3.3.4 Delivering Results to the Web

With the popularity of the Internet, you may find it useful to deploy selected pieces of output on your intranet or Web site. ODS makes deploying output to the Web a snap. The HTML destination creates syntactically correct HTML code to be used with one of the leading Internet browsers.

Four types of files can be produced with the ODS HTML destination: 1) body, 2) contents, 3) page, and 4) frame. A unique file name must be assigned to each file created with the ODS HTML statement. A custom and integrated file structure is automatically created when each file is combined with a frame file. To improve navigation and access of information, the Web browser automatically places horizontal and vertical scroll bars on the generated page, if necessary.

The next example illustrates PROC SQL output being sent to external HTML files using the HTML format engine. First, the default Listing destination is closed, and then the HTML format engine is opened specifying BODY, CONTENTS, PAGE, and FRAME external files for the routing of SQL procedure results. Once the SQL procedure executes, the HTML destination is closed and the default Listing destination is opened.

SQL Code

```
ODS LISTING CLOSE;
ODS HTML      BODY='c:\Products-body.html'
         CONTENTS='c:\Products-contents.html'
             PAGE='c:\Products-page.html'
            FRAME='c:\Products-frame.html';
PROC SQL;
  TITLE1 'Products List';
  SELECT prodname, prodtype, prodcost, prodnum
    FROM PRODUCTS
      ORDER BY prodtype;
QUIT;
ODS HTML CLOSE;
ODS LISTING;
```

Results

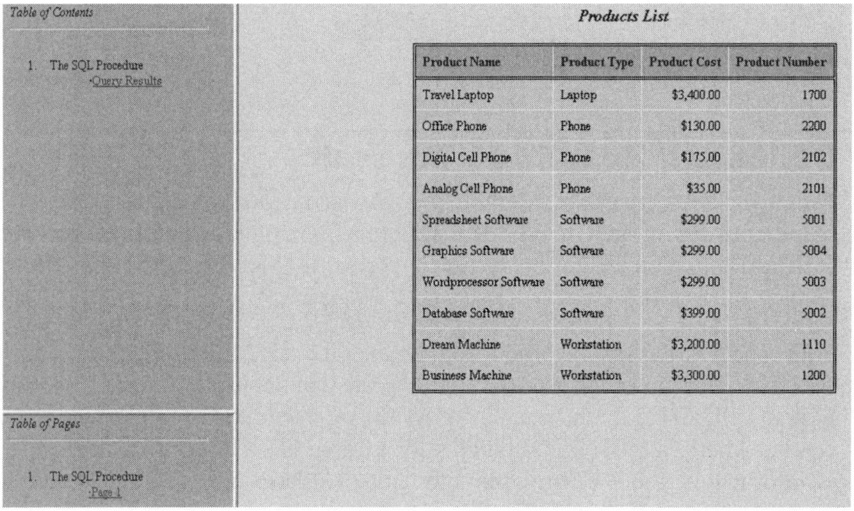

Table of Contents

1. The SQL Procedure
   - Query Results

Table of Pages

1. The SQL Procedure
   - Page 1

Products List

Product Name	Product Type	Product Cost	Product Number
Travel Laptop	Laptop	$3,400.00	1700
Office Phone	Phone	$130.00	2200
Digital Cell Phone	Phone	$175.00	2102
Analog Cell Phone	Phone	$35.00	2101
Spreadsheet Software	Software	$299.00	5001
Graphics Software	Software	$299.00	5004
Wordprocessor Software	Software	$299.00	5003
Database Software	Software	$399.00	5002
Dream Machine	Workstation	$3,200.00	1110
Business Machine	Workstation	$3,300.00	1200

3.4 Summary

1. A blank line can be displayed between each row of output (see section 3.2.1).
2. Columns can be concatenated to form a single column of data (see section 3.2.3).
3. Text and constants can be inserted between selected columns (see section 3.2.4).
4. Numeric or character scalar values can be produced with expressions (see section 3.2.5).
5. User-defined values can be assigned to derived column headers (see section 3.2.5).
6. Formats can be assigned and stored permanently to automatically display a user-defined formatted value instead of the unformatted value (see section 3.2.5).
7. Columns do not have to appear as unordered sets of data. One or more columns can be arranged in ascending or descending order (see section 3.2.6).
8. Selected columns can be organized and displayed in groups (see section 3.2.7).
9. PROC SQL can be coupled with ODS to extend output formatting capabilities (see section 3.3.1).

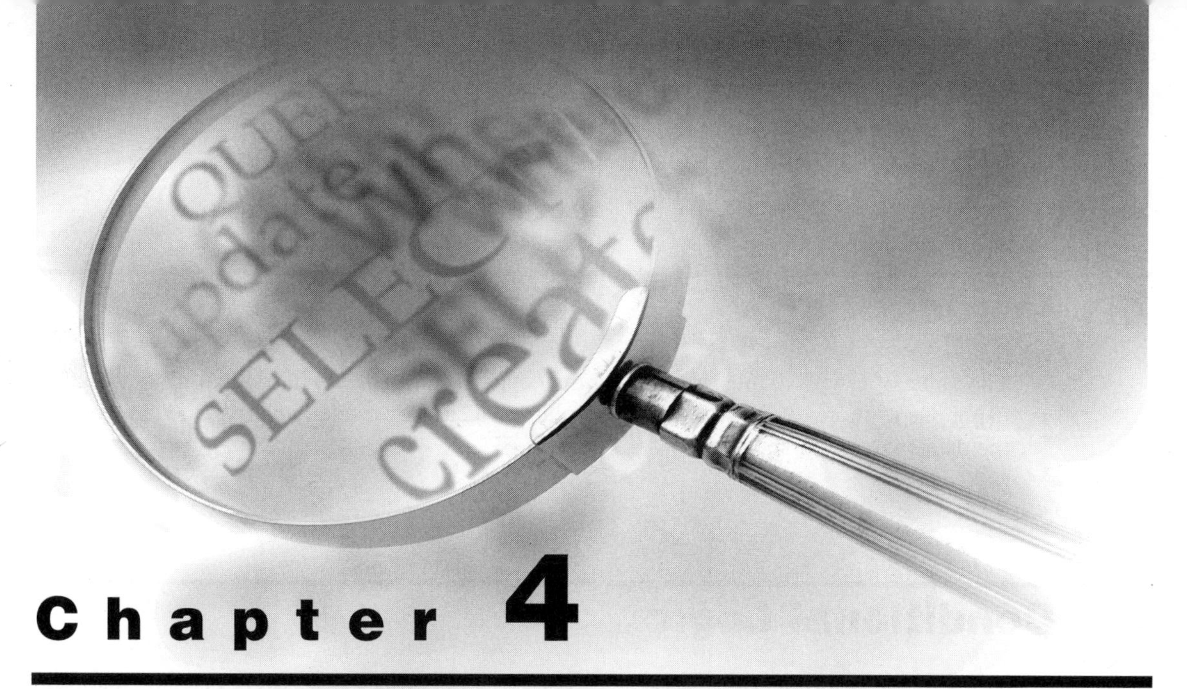

Chapter 4

Coding PROC SQL Logic

4.1 Introduction

The value of CASE expressions is that they enable you to group individual rows of data using one or more expressions. In particular, data can be recoded and reshaped to expand the data analysis perspective.

4.2 Conditional Logic

As experienced PROC SQL programmers know, it is often necessary to test and evaluate conditions as true or false. From a programming perspective, the evaluation of a condition determines which of the alternate paths a program will follow. Conditional logic in PROC SQL is frequently implemented in a WHERE clause to reference constants and relationships among columns and values. The SQL procedure also allows the identification and assignment of data values in a SELECT statement using CASE expressions (described in the next section). To show how constants and relationships are referenced in a WHERE clause, a simple example is given.

The WHERE clause expression evaluates whether the cost of a product (PRODCOST) is less than $400.00. During execution the expression evaluates to true only when the value of PRODCOST is less than 400.00. Otherwise, when the value is greater than or equal to 400.00, the expression evaluates to false. This is an important concept because any data rows satisfying the condition are only selected if the specified condition is true.

SQL Code

```
PROC SQL;
  SELECT *
    FROM PRODUCTS
      WHERE PRODCOST < 400.00;
QUIT;
```

The following relation tests whether the cost of a product (PRODCOST) is greater than $400.00. During execution the expression evaluates to true when the value of PRODCOST is greater than 400.00 and the rows of data satisfying the expression are selected. Otherwise, if the value is less than or equal to 400.00, the expression evaluates to false.

SQL Code

```
PROC SQL;
  SELECT *
    FROM PRODUCTS
      WHERE PRODCOST > 400.00;
QUIT;
```

A relation can also be used on nonnumeric literals and nonnumeric columns. In the next example, a case-sensitive expression is constructed to represent the type of product (PRODTYPE) made by a manufacturer. When evaluated, a condition of "true" or "false" is produced depending on whether the current value of PRODTYPE is identical (character-by-character) to the literal value "Software". When a condition of "true" occurs then the rows of data satisfying the expression are selected; otherwise they are not.

SQL Code

```
PROC SQL;
  SELECT *
    FROM PRODUCTS
      WHERE PRODTYPE = "Software";
QUIT;
```

To ensure a character-by-character match of a character value, the previous expression could be specified with the UPCASE function as follows.

SQL Code

```
PROC SQL;
  SELECT *
    FROM PRODUCTS
      WHERE UPCASE(PRODTYPE) = "SOFTWARE";
QUIT;
```

When the relations < and > are defined for nonnumeric values, the issue of implementation-dependent collating sequence for characters comes into play. For example, "A" < "B" is true, "Y" < "Z" is true, "B" < "A" is false, and so forth. For more information about character collating sequences, refer to the specific operating system documentation for the platform you are using.

4.3 CASE Expressions

In the SQL procedure, a CASE expression provides a way of determining what the resulting value will be from all rows in a table (or view). Similar to a DATA step SELECT statement (or IF-THEN/ELSE statement), a CASE expression is based on some condition and the condition uses a WHEN-THEN clause to determine what the resulting value will be. An optional ELSE expression can be specified to handle an alternative action if none of the expression(s) identified in the when-condition(s) is satisfied.

A CASE expression must be a valid PROC SQL expression and conform to syntax rules similar to DATA step SELECT-WHEN statements. Even though this topic is best explained by example, let's take a quick look at the syntax.

```
CASE <column-name>
     WHEN when-condition THEN result-expression
     <WHEN when-condition THEN result-expression> …
     <ELSE result-expression>
END
```

A column name can optionally be specified as part of the CASE expression. If present, it is automatically made available to each WHEN-condition. When it is not specified, the column name must be coded in each WHEN-condition. Let's examine how a CASE expression works.

If a WHEN-condition is satisfied by a row in a table (or view), then it is considered "true" and the result expression following the THEN expression is processed. The remaining WHEN conditions in the CASE expression are skipped. If a WHEN-condition is "false," the next WHEN-condition is evaluated. SQL evaluates each WHEN-condition until a "true" condition is found; or in the event all WHEN-conditions are "false," it then executes the ELSE expression and assigns its value to the CASE expression's result. A missing value is assigned to a CASE expression when an ELSE expression is not specified and each WHEN-condition is "false."

In the next example, let's see how a CASE expression actually works. Suppose a value of "West", "East", "Central", or "Unknown" is desired for each of the manufacturers. Using the manufacturer's state of residence (MANUSTAT) column, a CASE expression could be constructed to assign the desired value in a unique column for each row of data. A value of "West" is assigned to the manufacturers in California, "East" for manufacturers in Florida, "Central" for manufacturers in Texas, and for all other manufacturers a value of "Unknown" is assigned to represent missing state values. A column heading of Region is assigned to the new derived output column using the AS keyword.

SQL Code

```
PROC SQL;
  SELECT MANUNAME,
         MANUSTAT,
         CASE
           WHEN MANUSTAT = 'CA' THEN 'West'
           WHEN MANUSTAT = 'FL' THEN 'East'
           WHEN MANUSTAT = 'TX' THEN 'Central'
           ELSE 'Unknown'
         END AS Region
    FROM MANUFACTURERS;
QUIT;
```

Results

The SAS System

Manufacturer Name	Manufacturer State	Region
Cupid Computer	TX	Central
Global Comm Corp	CA	West
World Internet Corp	FL	East
Storage Devices Inc	CA	West
KPL Enterprises	CA	West
San Diego PC Planet	CA	West

Let's look at another example. In the PRODUCTS table a column called PRODTYPE contains the type of product (for example, Laptop, Phone, Software, and Workstation) as a text string. Using a CASE expression the assignment of a new data value is derived from the coded values in the PRODTYPE column: "Laptop" = " Hardware", " Phone" = "Hardware", "Software" = "Software", and "Workstation" = "Hardware". A column heading of Product_Classification is assigned to the derived output column with the AS keyword.

SQL Code

```
PROC SQL;
  SELECT PRODNAME,
         CASE PRODTYPE
           WHEN 'Laptop'      THEN 'Hardware'
           WHEN 'Phone'       THEN 'Hardware'
           WHEN 'Software'    THEN 'Software'
           WHEN 'Workstation' THEN 'Hardware'
           ELSE 'Unknown'
         END AS Product_Classification
    FROM PRODUCTS;
QUIT;
```

Results

```
                    The SAS System

                                      Product_
    Product Name                      Classification
    ________________________________________________

    Dream Machine                     Hardware
    Business Machine                  Hardware
    Travel Laptop                     Hardware
    Analog Cell Phone                 Hardware
    Digital Cell Phone                Hardware
    Office Phone                      Hardware
    Spreadsheet Software              Software
    Database Software                 Software
    Wordprocessor Software            Software
    Graphics Software                 Software
```

4.3.1 Case Logic versus COALESCE Expression

A popular convention among SQL users is to specify a COALESCE function in an expression to perform case logic. As described in Chapter 2, "Working with Data in PROC SQL," the COALESCE function permits a new value to be substituted for one or more missing column values. When you specify COALESCE in an expression, PROC SQL evaluates each argument from left to right for the occurrence of a nonmissing value.

The first nonmissing value found in the list of arguments is returned; otherwise a missing value, or assigned value, is returned. This approach not only saves programming time; it makes code simpler to maintain.

Expressing logical expressions in one or more WHEN-THEN/ELSE statements is generally easy, and the expressions are easy to code, understand, and maintain. But as the complexities associated with case logic increase, the amount of coding also increases. In the following example a simple CASE expression is presented to illustrate how a value of “Unknown” is assigned and displayed when CUSTCITY is missing.

SQL Case Logic

```
PROC SQL;
  SELECT CUSTNAME,
         CASE
           WHEN CUSTCITY IS NOT NULL THEN CUSTCITY
           ELSE 'Unknown'
         END AS Customer_City
    FROM CUSTOMERS;
QUIT;
```

To illustrate the usefulness of the COALESCE function as an alternative to case logic, the same query can be modified to achieve the same results as before. By replacing the case logic with a COALESCE expression as follows, the value of CUSTCITY is automatically displayed unless it is missing. In cases of character data a value of “Unknown” is displayed. This technique makes the COALESCE function a very useful and shorthand approach indeed.

SQL COALESCE Logic

```
PROC SQL;
  SELECT CUSTNAME,
         COALESCE(CUSTCITY,'Unknown')
           AS Customer_City
    FROM CUSTOMER;
QUIT;
```

In cases where a COALESCE expression is used with numeric data, the value assigned or displayed must be of the same type as the expression. The next example shows a value of “0” (zero) being assigned and displayed when units (UNITS) from the PURCHASES table are processed.

SQL COALESCE Logic

```
PROC SQL;
  SELECT ITEM,
         COALESCE(UNITS, 0)
    FROM PURCHASES;
QUIT;
```

Results

```
                 The SAS System

          item                   Units
          ____________________________

          Chair                      1
          Pens                      12
          Paper                      4
          Stapler                    1
          Mouse Pad                  1
          Pens                      24
          Markers                    0
```

4.3.2 Assigning Labels and Grouping Data

Assigning data values and group data based on the existence of distinct values for specified table columns is a popular and frequently useful operation. Suppose you want to assign a specific data value and then group the output based on this assigned value. As a savvy SAS user you are probably thinking, "Hey, this is easy — I'll just create a user-defined format and use it in the PRINT or REPORT procedure."

In the next example, PROC FORMAT is used to assign temporary formatted values based on a range of values for INVENQTY. The result from executing this simple three-step (non-SQL procedure) program shows that the actual INVENQTY value is temporarily replaced with the "matched" value in the user-defined format. The FORMAT statement performs a look-up process to determine how the data should be displayed. The actual data value being looked up is not changed during the process, but a determination is made as to how its value should be displayed. The BY statement specifies how BY-group processing is to be constructed. The displayed results show the product numbers in relation to their respective inventory quantity status.

Non-SQL Code

```
PROC FORMAT;
  VALUE INVQTY
       0 -   5 = 'Low on Stock - Reorder'
       6 -  10 = 'Stock Levels OK'
      11 -  99 = 'Plenty of Stock'
     100 - 999 = 'Excessive Quantities';
RUN;

PROC SORT DATA=INVENTORY;
  BY INVENQTY;
RUN;

PROC PRINT DATA=INVENTORY(KEEP=PRODNUM INVENQTY) NOOBS;
  FORMAT INVENQTY INVQTY.;
RUN;
```

Results

```
                  The SAS System

      prodnum              invenqty

        5001      Low on Stock - Reorder
        5002      Low on Stock - Reorder
        5001      Low on Stock - Reorder
        1700      Stock Levels OK
        5003      Stock Levels OK
        1110      Plenty of Stock
        5004      Plenty of Stock
```

The same results can also be derived using a CASE expression in the SQL procedure. In the next example, a CASE expression is constructed using the INVENTORY table to assign values to the user-defined column Inventory_Status. The biggest difference between the FORMAT procedure approach and a CASE expression is that the latter uses one step and does not replace the actual data value with the recoded result. Instead, it creates a new column containing the result of the CASE expression.

SQL Code

```
PROC SQL;
  SELECT PRODNUM,
         CASE
           WHEN INVENQTY LE 5
                THEN 'Low on Stock - Reorder'
           WHEN 6 LE INVENQTY LE 10
                THEN 'Stock Levels OK'
           WHEN 11 LE INVENQTY LE 99
                THEN 'Plenty of Stock'
           ELSE 'Excessive Quantities'
         END AS Inventory_Status
    FROM INVENTORY
      ORDER BY INVENQTY;
QUIT;
```

Results

```
                    The SAS System

        Product
         Number    Inventory_Status
        ________________________________
           5001    Low on Stock - Reorder
           5002    Low on Stock - Reorder
           5001    Low on Stock - Reorder
           1700    Stock Levels OK
           5003    Stock Levels OK
           1110    Plenty of Stock
           5004    Plenty of Stock
```

4.3.3 Logic and Nulls

The existence of null values frequently introduces complexities for programmers. Instead of coding two-valued logic conditions, such as true and false, logic conditions must be designed to handle three-valued logic: true, false, and unknown. When developing logic conditions, you need to be ready to deal with the possibility of having null values. Program logic should test whether the current value of an expression contains a value or is empty (null).

Let's examine a CASE expression that is meant to handle the possibility of having missing values in a table. Returning to an example presented earlier in this chapter, suppose we want to assign a value of "South West", "South East", "Central", "Missing", or "Unknown" to each of the manufacturers based on their state of residence.

SQL Code

```
PROC SQL;
  SELECT MANUNAME,
         MANUSTAT,
         CASE
           WHEN MANUSTAT = 'CA' THEN 'South West'
           WHEN MANUSTAT = 'FL' THEN 'South East'
           WHEN MANUSTAT = 'TX' THEN 'Central'
           WHEN MANUSTAT = ' '  THEN 'Missing'
           ELSE 'Unknown'
         END AS Region
    FROM MANUFACTURERS;
QUIT;
```

Results

The SAS System

Manufacturer Name	Manufacturer State	Region
Cupid Computer	TX	Central
Global Comm Corp	CA	South West
World Internet Corp	FL	South East
Storage Devices Inc	CA	South West
KPL Enterprises	CA	South West
San Diego PC Planet	CA	South West

The results indicate that there were no missing or null values in our database for the column being tested. But, suppose a new row of data were added containing null values in the manufacturer's city and state of residence columns so our new row looked something like the following:

```
Manufacturer Number: 800
Manufacturer Name: Spring Valley Products
Manufacturer City: <Missing>
Manufacturer State: <Missing>.
```

The result would look something like the following if we ran the previous code again.

SQL Code

```
PROC SQL;
  SELECT MANUNAME,
         MANUSTAT,
         CASE
           WHEN MANUSTAT = 'CA' THEN 'South West'
           WHEN MANUSTAT = 'FL' THEN 'South East'
           WHEN MANUSTAT = 'TX' THEN 'Central'
           WHEN MANUSTAT = ' '  THEN 'Missing'
           ELSE 'Unknown'
         END AS Region
    FROM MANUFACTURERS;
QUIT;
```

Results

The SAS System

Manufacturer Name	Manufacturer State	Region
Cupid Computer	TX	Central
Global Comm Corp	CA	South West
World Internet Corp	FL	South East
Storage Devices Inc	CA	South West
KPL Enterprises	CA	South West
San Diego PC Planet	CA	South West
Spring Valley Products		Missing

4.4 Interfacing PROC SQL with the Macro Language

Many software vendors' SQL implementations permit SQL to be interfaced with a host language. The SAS System's SQL implementation is no different. The SAS macro language lets you customize the way SAS software behaves and in particular extend the capabilities of the SQL procedure. PROC SQL users can apply the macro facility's many powerful features by interfacing the SQL procedure with the macro language to provide a wealth of programming opportunities.

From creating and using user-defined macro variables and automatic (SAS-supplied) variables, reducing redundant code, performing common and repetitive tasks, to building powerful and simple macro applications, the macro language has the tools PROC SQL users can take advantage of to improve efficiency. The best part is that you do not have to be a macro language heavyweight to begin reaping the rewards of this versatile interface between two powerful Base SAS software languages.

This section will introduce you to a number of techniques that, with a little modification, could be replicated and used in your own programming environment. You will learn how to use the SQL procedure with macro programming techniques, as well as explore how dictionary tables (see Chapter 2 for details) and the SAS macro facility can be combined with PROC SQL to develop useful utilities to inquire about the operating environment and other information. For more information about the SAS Macro Language, readers are referred to *Carpenter's Complete Guide to the SAS Macro Language* by Art Carpenter; *SAS Macro Programming Made Easy* by Michele M. Burlew; and *SAS Macro Language: Reference, Version 8.*

4.4.1 Exploring Macro Variables and Values

Macro variables and their values provide PROC SQL users with a convenient way to store text strings in SAS code. Whether you create your own macro variables or use SAS-supplied automatic macro variables, macro variables can improve a program's efficiency and usefulness. A number of techniques are presented in this section to illustrate the capabilities afforded users when interfacing PROC SQL with macro variables.

4.4.1.1 Creating a Macro Variable with %LET

The %LET macro statement creates a single macro variable and assigns or changes a text string value. It can be specified inside or outside a macro and used with PROC SQL. In the next example, a macro variable called PRODTYPE is created with a value of SOFTWARE assigned in a %LET statement. The PRODTYPE macro variable is referenced in the TITLE statement and enclosed in quotes in the PROC SQL WHERE clause. This approach of assigning macro variable values at the beginning of a program makes it easy and convenient for making changes, if necessary, because the values are all at the beginning of the program.

SQL Code

```
%LET PRODTYPE=SOFTWARE;
TITLE "Listing of &PRODTYPE Products";
PROC SQL;
  SELECT PRODNAME,
         PRODCOST
    FROM PRODUCTS
      WHERE UPCASE(PRODTYPE) = "&PRODTYPE"
        ORDER BY PRODCOST;
QUIT;
```

Results

```
             Listing of SOFTWARE Products

                                          Product
         Product Name                        Cost
         ----------------------------------------

         Wordprocessor Software           $299.00
         Spreadsheet Software             $299.00
         Graphics Software                $299.00
         Database Software                $399.00
```

In the next example, a macro named VIEW creates a macro variable called NAME and assigns a value to it with a %LET statement. When VIEW is executed, a value of PRODUCTS, MANUFACTURERS, or INVENTORY is substituted for the macro variable. The value supplied for the macro variable determines what view is referenced. If the value supplied to the macro variable is not one of these three values, then a program warning message is displayed in the SAS log. Invoking the macro with %VIEW(Products) produces the following results.

SQL Code

```
%MACRO VIEW(NAME);
%IF %UPCASE(&NAME) ^= %STR(PRODUCTS) AND
    %UPCASE(&NAME) ^= %STR(MANUFACTURERS) AND
    %UPCASE(&NAME) ^= %STR(INVENTORY) %THEN %DO;
  %PUT A valid view name was not supplied and no output
       will be generated!;
%END;
%ELSE %DO;
  PROC SQL;
  TITLE "Listing of &NAME View";
  %IF %UPCASE(&NAME)=%STR(PRODUCTS) %THEN %DO;
    SELECT PRODNAME,
           PRODCOST
      FROM &NAME._view
        ORDER BY PRODCOST;
  %END;
  %ELSE %IF %UPCASE(&NAME)=%STR(MANUFACTURERS) %THEN %DO;
    SELECT MANUNAME,
           MANUCITY,
           MANUSTAT
      FROM &NAME._view
        ORDER BY MANUCITY;
  %END;
  %ELSE %IF %UPCASE(&NAME)=%STR(INVENTORY) %THEN %DO;
    SELECT PRODNUM,
           INVENQTY,
           INVENCST
      FROM &NAME._view
        ORDER BY INVENCST;
  %END;
  QUIT;
  %END;
%MEND VIEW;
```

Results

```
              Listing of Products View

                                               Product
           Product Name                           Cost
           ____________________________________________

           Analog Cell Phone                    $35.00
           Office Phone                        $130.00
           Digital Cell Phone                  $175.00
           Spreadsheet Software                $299.00
           Graphics Software                   $299.00
           Wordprocessor Software              $299.00
           Database Software                   $399.00
           Dream Machine                     $3,200.00
           Business Machine                  $3,300.00
           Travel Laptop                     $3,400.00
```

In the previous example, if a name is supplied to the macro variable &NAME that is not valid, the user-defined program warning message appears in the SAS log. Suppose we invoked the VIEW macro by entering %VIEW(Customers).

SQL Code

```
%VIEW(Customers);
```

SAS Log Results

```
  %VIEW(Customers);
A valid view name was not supplied and no output will be generated!
```

4.4.1.2 Creating a Macro Variable from a Table Row Column

A macro variable can be created from a column value in the first row of a table in PROC SQL by specifying the INTO clause. The macro variable is assigned using the value of the column specified in the SELECT list from the first row selected. A colon (:) is used in conjunction with the macro variable name being defined. In the next example, output results are suppressed with the NOPRINT option, while two macro variables are created using the INTO clause and their values displayed in the SAS log.

SQL Code

```
PROC SQL NOPRINT;
  SELECT PRODNAME,
         PRODCOST
    INTO :PRODNAME,
         :PRODCOST
      FROM PRODUCTS;
QUIT;
%PUT &PRODNAME &PRODCOST;
```

SAS Log Results

```
PROC SQL NOPRINT;
  SELECT PRODNAME,
         PRODCOST
    INTO :PRODNAME,
         :PRODCOST
      FROM PRODUCTS;
QUIT;
NOTE: PROCEDURE SQL used:
      real time           0.38 seconds

%PUT &PRODNAME &PRODCOST;
Dream Machine              $3,200.00
```

In the next example, two macro variables are created using the INTO clause and a WHERE clause to control what row is used in the assignment of macro variable values. Using the WHERE clause enables a row other than the first row to always be used in the assignment of macro variables. Their values are displayed in the SAS log.

SQL Code

```
PROC SQL NOPRINT;
  SELECT PRODNAME,
         PRODCOST
    INTO :PRODNAME,
         :PRODCOST
      FROM PRODUCTS
        WHERE UPCASE(PRODTYPE) IN ('SOFTWARE');
QUIT;
%PUT &PRODNAME &PRODCOST;
```

SAS Log Results

```
PROC SQL NOPRINT;
  SELECT PRODNAME,
         PRODCOST
    INTO :PRODNAME,
         :PRODCOST
      FROM PRODUCTS
        WHERE UPCASE(PRODTYPE) IN ('SOFTWARE');
QUIT;
NOTE: PROCEDURE SQL used:
      real time           0.04 seconds

%PUT &PRODNAME &PRODCOST;
Spreadsheet Software        $299.00
```

4.4.1.3 Creating a Macro Variable with Aggregate Functions

Turning data into information and then saving the results as macro variables is easy with aggregate (summary) functions. The SQL procedure provides a number of useful aggregrate functions to help perform calculations, descriptive statistics, and other aggregating computations in a SELECT statement or HAVING clause. These functions are designed to summarize information and not display detail about data. In the next example, the MIN summary function is used to determine the least expensive product from the PRODUCTS table with the value stored in the macro variable MIN_PRODCOST using the INTO clause. The results are displayed in the SAS log.

SQL Code

```
PROC SQL NOPRINT;
  SELECT MIN(PRODCOST) FORMAT=DOLLAR10.2
    INTO :MIN_PRODCOST
      FROM PRODUCTS;
QUIT;
%PUT &MIN_PRODCOST;
```

SAS Log Results

```
PROC SQL NOPRINT;
  SELECT MIN(PRODCOST) FORMAT=DOLLAR10.2
    INTO :MIN_PRODCOST
      FROM SQL.PRODUCTS;
QUIT;
NOTE: PROCEDURE SQL used:
      real time           0.05 seconds

%PUT &MIN_PRODCOST;
$35.00
```

4.4.1.4 Creating Multiple Macro Variables

PROC SQL lets you create a macro variable for each row returned by a SELECT statement. Using the PROC SQL keyword THROUGH or hyphen (-) with the INTO clause, you can easily create a range of two or more macro variables. This is a handy feature for creating macro variables from multiple rows in a table. For example, suppose we wanted to create macro variables for the three least expensive products in the PRODUCTS table. The INTO clause creates three macro variables and assigns values from the first three rows of the PRODNAME and PRODCOST columns. The ORDER BY clause is also specified to perform an ascending sort on product cost (PRODCOST) to assure that the data is in the desired order from least to most expensive. The results are displayed on the SAS log.

SQL Code

```
PROC SQL NOPRINT;
  SELECT PRODNAME,
         PRODCOST
    INTO :PRODUCT1 - :PRODUCT3,
         :COST1 - :COST3
      FROM PRODUCTS
        ORDER BY PRODCOST;
QUIT;
%PUT &PRODUCT1 &COST1;
%PUT &PRODUCT2 &COST2;
%PUT &PRODUCT3 &COST3;
```

SAS Log Results

```
PROC SQL NOPRINT;
  SELECT PRODNAME,
         PRODCOST
    INTO :PRODUCT1 - :PRODUCT3,
         :COST1 - :COST3
      FROM PRODUCTS
        ORDER BY PRODCOST;
QUIT;
NOTE: PROCEDURE SQL used:
      real time           0.26 seconds

%PUT &PRODUCT1 &COST1;
Analog Cell Phone $35.00
%PUT &PRODUCT2 &COST2;
Office Phone $130.00
%PUT &PRODUCT3 &COST3;
Digital Cell Phone $175.00
```

4.4.1.5 Creating a List of Values in a Macro Variable

Concatenating values of a single column into one macro variable lets you create a list of values that can be displayed in the SAS log or output to a SAS data set. Use the INTO clause with the SEPARATED BY option to create a list of values. For example, suppose we wanted to create a blank-delimited list containing manufacturer names (MANUNAME) from the MANUFACTURERS table. We create a macro variable called

&MANUNAME and assign the manufacturer names to a blank-delimited list with each name separated with two blank spaces. The WHERE clause restricts the list's contents to only those manufacturers located in San Diego.

SQL Code

```
PROC SQL NOPRINT;
  SELECT MANUNAME
    INTO :MANUNAME SEPARATED BY '  '
      FROM MANUFACTURERS
        WHERE UPCASE(MANUCITY)='SAN DIEGO';
QUIT;
%PUT &MANUNAME;
```

SAS Log Results

```
PROC SQL NOPRINT;
  SELECT MANUNAME
    INTO :MANUNAME SEPARATED BY '  '
      FROM MANUFACTURERS
        WHERE UPCASE(MANUCITY)='SAN DIEGO';
QUIT;
NOTE: PROCEDURE SQL used:
      real time           0.00 seconds

%PUT &MANUNAME;
Global Comm Corp  KPL Enterprises  San Diego PC Planet
```

In the next example, a similar list containing manufacturers from San Diego is created. But instead of each name being separated with two blanks as in the previous example, the names are separated by commas.

SQL Code

```
PROC SQL NOPRINT;
  SELECT MANUNAME
    INTO :MANUNAME SEPARATED BY ', '
      FROM MANUFACTURERS
        WHERE UPCASE(MANUCITY)='SAN DIEGO';
QUIT;
%PUT &MANUNAME;
```

SAS Log Results

```
PROC SQL NOPRINT;
  SELECT MANUNAME
    INTO :MANUNAME SEPARATED BY ', '
      FROM MANUFACTURERS
        WHERE UPCASE(MANUCITY)='SAN DIEGO';
QUIT;
NOTE: PROCEDURE SQL used:
      real time           0.00 seconds

%PUT &MANUNAME;
Global Comm Corp, KPL Enterprises, San Diego PC Planet
```

4.4.1.6 Using Automatic Macro Variables to Control Processing

Three automatic macro variables supplied by SAS are assigned values during SQL processing for the purpose of providing process control information. SQL users can determine the number of rows processed with the SQLOBS macro variable, assess whether a PROC SQL statement was successful or not with the SQLRC macro variable, or identify the number of iterations the inner loop of an SQL query processes with the SQLOOPS macro variable. To inspect the values of all automatic macro variables at your installation, use the _AUTOMATIC_ option in a %PUT statement.

SQL Code

```
%PUT _AUTOMATIC_;
```

SAS Log Results

```
%PUT _AUTOMATIC_;
AUTOMATIC AFDSID 0
AUTOMATIC AFDSNAME
AUTOMATIC AFLIB
AUTOMATIC AFSTR1
AUTOMATIC AFSTR2
```

(continued on next page)

```
AUTOMATIC FSPBDV
AUTOMATIC SYSBUFFR
AUTOMATIC SYSCC 0
AUTOMATIC SYSCHARWIDTH 1
AUTOMATIC SYSCMD
AUTOMATIC SYSDATE 10JUN04
AUTOMATIC SYSDATE9 10JUN2004
AUTOMATIC SYSDAY Thursday
AUTOMATIC SYSDEVIC
AUTOMATIC SYSDMG 0
AUTOMATIC SYSDSN WORK    INVENTORY
AUTOMATIC SYSENDIAN LITTLE
AUTOMATIC SYSENV FORE
AUTOMATIC SYSERR 0
AUTOMATIC SYSFILRC 0
AUTOMATIC SYSINDEX 3
AUTOMATIC SYSINFO 0
AUTOMATIC SYSJOBID 3580
AUTOMATIC SYSLAST WORK.INVENTORY
AUTOMATIC SYSLCKRC 0
AUTOMATIC SYSLIBRC 0
AUTOMATIC SYSMACRONAME
AUTOMATIC SYSMAXLONG 2147483647
AUTOMATIC SYSMENV S
AUTOMATIC SYSMSG
AUTOMATIC SYSNCPU 1
AUTOMATIC SYSPARM
AUTOMATIC SYSPBUFF
AUTOMATIC SYSPROCESSID 41D4E614295031274020000000000000
AUTOMATIC SYSPROCESSNAME DMS Process
AUTOMATIC SYSPROCNAME
AUTOMATIC SYSRC 0
AUTOMATIC SYSSCP WIN
AUTOMATIC SYSSCPL XP_HOME
AUTOMATIC SYSSITE 0045254001
AUTOMATIC SYSSIZEOFLONG 4
AUTOMATIC SYSSIZEOFUNICODE 2
```

(continued on next page)

```
AUTOMATIC SYSSTARTID
AUTOMATIC SYSSTARTNAME
AUTOMATIC SYSTIME 12:50
AUTOMATIC SYSUSERID Valued Sony Customer
AUTOMATIC SYSVER 9.1
AUTOMATIC SYSVLONG 9.01.01M0P111803
AUTOMATIC SYSVLONG4 9.01.01M0P11182003
```

4.4.2 Building Macro Tools and Applications

The macro facility, combined with the capabilities of the SQL procedure, enables the creation of versatile macro tools and general-purpose applications. A principal design goal when writing macros is that they are useful and simple to use. A macro that has little applicability to user needs or has complicated and hard to remember macro variable names is best avoided.

As tools, macros should be designed to serve the needs of as many users as possible. They should contain no ambiguities, consist of distinctive macro variable names, avoid the possibility of naming conflicts between macro variables and data set variables, and not try to do too many things. This utilitarian approach to macro design helps gain widespread approval and acceptance by users.

4.4.2.1 Creating Simple Macro Tools

Macro tools can be constructed to perform a variety of useful tasks. The most effective ones are usually those that are simple and perform a common task. Before constructing one or more macro tools, explore what processes are currently being performed, then identify common users' needs with affected personnel by addressing gaps in the process. Once this has been accomplished, you will be in a better position to construct simple and useful macro tools that will be accepted by user personnel.

Suppose during an informal requirements analysis phase that you identified users who, in the course of their jobs, use a variety of approaches and methods to create data set and variable cross-reference listings. To prevent unnecessary and wasteful duplication of effort, you decide to construct a simple macro tool that can be used by all users to retrieve information about the columns in one or more SAS data sets.

4.4.2.2 Cross-Referencing Columns

Column cross-reference listings come in handy when you need to quickly identify all the SAS library data sets a column is defined in. Using the COLUMNS dictionary table (see Chapter 2, "Working with Data in PROC SQL," for more details) a macro can be created that captures column-level information including column name, type, length, position, label, format, informat, indexes, as well as a cross-reference listing containing the location of a column within a designated SAS library. In the next example, macro COLUMNS consists of a PROC SQL query that accesses any single column in a SAS library. If the macro was invoked with a user-request consisting of %COLUMNS(WORK,CUSTNUM), the macro would produce a cross-reference listing on the user library WORK for the column CUSTNUM in all DATA types.

SQL Code

```
%MACRO COLUMNS(LIB, COLNAME);
  PROC SQL;
    SELECT LIBNAME, MEMNAME, NAME, TYPE, LENGTH
      FROM DICTIONARY.COLUMNS
        WHERE LIBNAME='&LIB' AND
              UPCASE(NAME)='&COLNAME' AND
              MEMTYPE='DATA';
  QUIT;
%MEND COLUMNS;


%COLUMNS(WORK,CUSTNUM);
```

It is worth noting that multiple matches could be found in databases containing case-sensitive names. This would allow both "employee" and "EMPLOYEE" to be displayed as matches. This is not very likely to occur too often in practice but is definitely a possibility.

Results

Library Name	Member Name	Column Name	Column Type	Column Length
WORK	CUSTOMERS	custnum	num	3
WORK	INVOICE	custnum	num	3
WORK	PURCHASES	custnum	num	4

4.4.2.3 Determining the Number of Rows in a Table

Sometimes it is useful to know the number of observations (or rows) in a table without first having to read all the rows. Although the number of rows in a table is available for true SAS System tables, they are not for DBMS tables using a library engine. In the next example, the TABLES dictionary table is accessed (refer to Chapter 2 for more details) in a user-defined macro called NOBS. Macro NOBS is designed to accept and process two user-supplied values: the library reference and the table name. Once these values are supplied, the results are displayed in the Output window.

SQL Code

```
%MACRO NOBS(LIB, TABLE);
  PROC SQL;
    SELECT LIBNAME, MEMNAME, NOBS
      FROM DICTIONARY.TABLES
        WHERE UPCASE(LIBNAME)="&LIB" AND
              UPCASE(MEMNAME)="&TABLE" AND
              UPCASE(MEMTYPE)="DATA";
  QUIT;
%MEND NOBS;


%NOBS(WORK,PRODUCTS);
```

Results

```
                   The SAS System

Library                                   Number of
Name       Member Name                 Observations
---------------------------------------------------
WORK       PRODUCTS                              10
```

4.4.2.4 Identifying Duplicate Rows in a Table

Sometimes it is handy to be able to identify duplicate rows in a table. In the next example, the SELECT statement with a COUNT summary function and HAVING clause are used in a user-defined macro called DUPS. Macro DUPS is designed to accept and process three user-supplied values: the library reference, table name, and column(s) in a

GROUP BY list. Once these values are supplied by submitting macro DUPS, the macro is executed with the results displayed in the Output window.

SQL Code

```
%MACRO DUPS(LIB, TABLE, GROUPBY);
  PROC SQL;
    SELECT &GROUPBY, COUNT(*) AS Duplicate_Rows
      FROM &LIB..&TABLE
        GROUP BY &GROUPBY
          HAVING COUNT(*) > 1;
  QUIT;
%MEND DUPS;

%DUPS(WORK,PRODUCTS,PRODTYPE);
```

Results

```
                    The SAS System

                              Duplicate_
        Product Type                Rows
        ________________________________

        Phone                          3
        Software                       4
        Workstation                    2
```

4.5 Summary

1. A CASE expression is a PROC SQL feature that can be used to evaluate whether a particular condition has been met (see section 4.3).

2. A CASE expression can be used to process a table's rows (see section 4.3).

3. A single value is returned from its evaluation of each row in a table (or view) (see section 4.3).

4. Logic conditions can be combined using the logical operators AND and OR (see section 4.3.3).

5. A missing or null value is returned when an ELSE expression is not specified and each when-condition is “false” (see section 4.3.3).

6. A missing value is not the same as a value of 0 (zero) or as a blank character since it represents a unique value or a lack of a value (see section 4.3.3).

7. PROC SQL can be used with the SAS macro facility to perform common and repetitive tasks (see section 4.4).

8. Simple, but effective, user-defined macros combined with the SQL procedure can be created for all users (see section 4.4.2).

9. Identify duplicate rows in a table by creating a user-defined macro (see section 4.4.2.4).

Chapter 5

Creating, Populating, and Deleting Tables

5.1 Introduction

Previous chapters provided tables in the examples that had already been created and populated with data. But what if you need to create a table, populate it with data, or delete rows of data or tables that are no longer needed or wanted?

In this chapter, the numerous discussions and examples focus on the way tables are created, populated and deleted. These are important operations and essential elements in PROC SQL, especially if you want to increase your comprehension of SQL processes and improve your understanding of this powerful language.

5.2 Creating Tables

An important element in table creation is table design. Table design incorporates how tables are structured — how rows and columns are defined, how indexes are created, and how columns refer to values in other columns. Readers seeking a greater understanding in this area are encouraged to review the many references identified at the end of this book. The following overview should be kept in mind during the table design process.

When building a table it is important to devote adequate time to planning its design as well as understanding the needs that each table is meant to satisfy. This process involves a number of activities such as requirements and feasibility analysis including cost/benefit of the proposed tables, the development of a logical description of the data sources, and physical implementation of the logical data model. Once these tasks are complete, you assess any special business requirements that each table is to provide. A business assessment helps by minimizing the number of changes required to a table once it has been created.

Next, determine what tables will be incorporated into your application's database. This requires understanding the value that each table is expected to provide. It also prevents a table of little or no importance from being incorporated into a database. The final step and one of critical importance is to define each table's columns, attributes, and contents.

Once the table design process is complete, each table is then ready to be created with the CREATE TABLE statement. The purpose of creating a table is to create an object that does not already exist. In the SAS implementation, three variations of the CREATE TABLE statement can be specified depending on your needs:

- Creating a table structure with column-definition lists
- Creating a table structure with the LIKE clause
- Deriving a table structure and data from an existing table
- Creates new columns

5.2.1 Creating a Table Using Column-Definition Lists

Although part of the SQL standard, the column-definition list (like the LENGTH statement in the DATA step) is a laborious and not very elegant way to create a table. The disadvantage of creating a table this way is that it requires the definition of each column's attributes including their data type, length, informat, and format. This method is frequently used to create columns when they are not present in another table. Using this method results in the creation of an empty table *(without rows)*. The code used to create the CUSTOMERS table appears below. It illustrates the creation of a table with column-definition lists.

SQL Code

```
PROC SQL;
  CREATE TABLE CUSTOMERS
    (CUSTNUM   NUM       LABEL='Customer Number',
     CUSTNAME  CHAR(25) LABEL='Customer Name',
     CUSTCITY  CHAR(20) LABEL='Customer''s Home City');
QUIT;
```

SAS Log Results

```
     PROC SQL;
       CREATE TABLE CUSTOMERS
         (CUSTNUM   NUM       LABEL='Customer Number',
          CUSTNAME  CHAR(25) LABEL='Customer Name',
          CUSTCITY  CHAR(20) LABEL='Customer''s Home City');
NOTE: Table CUSTOMERS created, with 0 rows and 3 columns.
     QUIT;
NOTE: PROCEDURE SQL used:
     real time           0.81 seconds
```

Readers should be aware that the SQL procedure ignores width specifications for numeric columns. When a numeric column is defined, it is created with a width of 8 bytes, which is the maximum precision allowed by the SAS System. PROC SQL ignores numeric length specifications when the value is less than 8 bytes. To illustrate this point, a partial CONTENTS procedure output is displayed for the CUSTOMERS table below.

Results

The CONTENTS Procedure

-----Alphabetic List of Variables and Attributes-----

#	Variable	Type	Len	Pos	Label
3	CUSTCITY	Char	20	33	Customer's Home City
2	CUSTNAME	Char	25	8	Customer Name
1	**CUSTNUM**	**Num**	**8**	**0**	**Customer Number**

To conserve storage space (CUSTNUM only requires maximum precision provided in 3 bytes), a LENGTH statement could be used in a DATA step to define CUSTNUM as a 3-byte column rather than an 8-byte column. A DROP= data set option is specified to delete the original CUSTNUM column (created by the CREATE TABLE statement) in the Program Data Vector or PDV.

DATA Step Code

```
DATA CUSTOMERS;
  LENGTH CUSTNUM 3.;
  SET CUSTOMERS(DROP=CUSTNUM);
  LABEL CUSTNUM = 'Customer Number';
RUN;
```

Results

The CONTENTS Procedure

-----Alphabetic List of Variables and Attributes-----

#	Variable	Type	Len	Pos	Label
3	CUSTCITY	Char	20	25	Customer's Home City
2	CUSTNAME	Char	25	0	Customer Name
1	**CUSTNUM**	**Num**	**3**	**45**	**Customer Number**

Let's look at the column-definition list used to create the PRODUCTS table.

SQL Code

```
PROC SQL;
  CREATE TABLE PRODUCTS
     (PRODNUM   NUM(3)    LABEL=`Product Number',
      PRODNAME  CHAR(25) LABEL=`Product Name',
      MANUNUM   NUM(3)    LABEL=`Manufacturer Number',
      PRODTYPE  CHAR(15) LABEL=`Product Type',
      PRODCOST  NUM(5,2) FORMAT=DOLLAR9.2 LABEL=`Product Cost');
QUIT;
```

SAS Log Results

```
   PROC SQL;
     CREATE TABLE PRODUCTS
       (PRODNUM   NUM(3)    LABEL='Product Number',
        PRODNAME  CHAR(25) LABEL='Product Name',
        MANUNUM   NUM(3)    LABEL='Manufacturer Number',
        PRODTYPE  CHAR(15) LABEL='Product Type',
        PRODCOST  NUM(5,2) FORMAT=DOLLAR9.2 LABEL='Product Cost');
NOTE: Table PRODUCTS created, with 0 rows and 5 columns.
   QUIT;
NOTE: PROCEDURE SQL used:
      real time           0.00 seconds
```

The CONTENTS output for the PRODUCTS table shows once again that the SQL procedure ignores all width specifications for numeric columns.

Results

```
                      The CONTENTS Procedure

       -----Alphabetic List of Variables and Attributes-----

 #    Variable    Type    Len    Pos    Format       Label
 ___________________________________________________________________
 3    MANUNUM     Num       8      8                 Manufacturer Number
 5    PRODCOST    Num       8     16    DOLLAR9.2    Product Cost
 2    PRODNAME    Char     25     24                 Product Name
 1    PRODNUM     Num       8      0                 Product Number
 4    PRODTYPE    Char     15     49                 Product Type
```

As before, to conserve storage space you can use a LENGTH statement in a DATA step to override the default 8-byte column definition for numeric columns.

DATA Step Code

```
DATA PRODUCTS;
  LENGTH PRODNUM MANUNUM 3.
         PRODCOST 5.;
  SET PRODUCTS(DROP=PRODNUM MANUNUM PRODCOST);
  LABEL PRODNUM  = 'Product Number'
        MANUNUM  = 'Manufacturer Number'
        PRODCOST = 'Product Cost';
  FORMAT PRODCOST DOLLAR9.2;
RUN;
```

Results

```
                    The CONTENTS Procedure

    -----Alphabetic List of Variables and Attributes-----

#   Variable    Type    Len    Format        Label
------------------------------------------------------------------
2   MANUNUM     Num       3                  Manufacturer Number
3   PRODCOST    Num       5    DOLLAR9.2     Product Cost
1   PRODNUM     Num       3                  Product Number
4   PRODNAME    Char     25                  Product Name
5   PRODTYPE    Char     15                  Product Type
```

5.2.2 Creating a Table Using the LIKE Clause

Referencing an existing table in a CREATE TABLE statement is an effective way of creating a new table. In fact, it can be a great time-saver, because it prevents having to define each column one at a time as was shown with column-definition lists. The LIKE clause (in the CREATE TABLE statement) triggers the existing table's structure to be copied to the new table minus any columns dropped with the KEEP= or DROP= data set (table) option. It copies the column names and attributes from the existing table structure to the new table structure. Using this method results in the creation of an empty table *(without rows)*. To illustrate this method of creating a new table, a table called HOT_PRODUCTS will be created with the LIKE clause.

SQL Code

```
PROC SQL;
  CREATE TABLE HOT_PRODUCTS
    LIKE PRODUCTS;
QUIT;
```

SAS Log Results

```
     PROC SQL;
       CREATE TABLE HOT_PRODUCTS
         LIKE PRODUCTS;
NOTE: Table HOT_PRODUCTS created, with 0 rows and 5 columns.
     QUIT;
NOTE: PROCEDURE SQL used:
      real time           0.00 seconds
```

Readers are reminded that, as a result of executing the LIKE clause in the CREATE TABLE statement, only those columns in the existing table are copied to the new table. What this means is that the new table has zero rows of data.

Our next example illustrates how to create a new table by selecting just the columns you have an interest in. This method is not supported by the SQL ANSI standard. Suppose you want three columns (PRODNAME, PRODTYPE, and PRODCOST) from the PRODUCTS table. The following code illustrates how the KEEP= data set (table) option can be used to accomplish this. (Note that data sets can also be called tables.)

SQL Code

```
PROC SQL;
  CREATE TABLE HOT_PRODUCTS(KEEP=PRODNAME PRODTYPE PRODCOST)
    LIKE PRODUCTS;
QUIT;
```

SAS Log Results

```
     PROC SQL;
        CREATE TABLE HOT_PRODUCTS(KEEP=PRODNAME PRODTYPE PRODCOST)
        LIKE PRODUCTS;
NOTE: Table HOT_PRODUCTS created, with 0 rows and 3 columns.
     QUIT;
NOTE: PROCEDURE SQL used:
      real time           0.00 seconds
```

5.2.3 Deriving a Table and Data from an Existing Table

Deriving a new table from an existing table is by far the most popular and effective way to create a table. This method uses a query expression, and the results are stored in a new table instead of being displayed as SAS output. This method not only stores the column names and their attributes, but the rows of data that satisfies the query expression as well. The following example illustrates how a new table is created using a query expression.

SQL Code

```
PROC SQL;
  CREATE TABLE HOT_PRODUCTS AS
    SELECT *
      FROM PRODUCTS;
QUIT;
```

SAS Log Results

```
     PROC SQL;
       CREATE TABLE HOT_PRODUCTS AS
         SELECT *
           FROM PRODUCTS;
NOTE: Table WORK.HOT_PRODUCTS created, with 10 rows and 5 columns.
     QUIT;
NOTE: PROCEDURE SQL used:
      real time           0.00 seconds
```

Readers may notice after examining the SAS log in the previous example that the SELECT statement extracted all rows from the existing table (PRODUCTS) and copied them to the new table (HOT_PRODUCTS). In the absence of a WHERE clause, the resulting table (HOT_PRODUCTS) contains the identical number of rows as the parent table PRODUCTS.

The power of the CREATE TABLE statement, then, is in its ability to create a new table from an existing table. What is often overlooked in this definition is the CREATE TABLE statement's ability to form a subset of a parent table. More frequently than not, a new table represents a subset of its parent table. For this reason this method of creating a table is the most powerful and widely used. Suppose you want to create a new table called HOT_PRODUCTS containing a subset of "Software" and "Phones" product types. The following query-expression would accomplish this.

SQL Code

```
PROC SQL;
  CREATE TABLE HOT_PRODUCTS AS
    SELECT *
      FROM PRODUCTS
        WHERE UPCASE(PRODTYPE) IN ("SOFTWARE", "PHONE");
QUIT;
```

SAS Log Results

```
     PROC SQL;
       CREATE TABLE HOT_PRODUCTS AS
         SELECT *
           FROM PRODUCTS
             WHERE UPCASE(PRODTYPE) IN ("SOFTWARE", "PHONE");
NOTE: Table WORK.HOT_PRODUCTS created, with 7 rows and 5 columns.
     QUIT;
NOTE: PROCEDURE SQL used:
      real time           0.38 seconds
```

Let's look at another example. Suppose you want to create another table called NOT_SO_HOT_PRODUCTS containing a subset of everything but "Software" and "Phones" product types. The following query-expression would accomplish this.

SQL Code

```
PROC SQL;
  CREATE TABLE NOT_SO_HOT_PRODUCTS AS
    SELECT *
      FROM PRODUCTS
        WHERE UPCASE(PRODTYPE) NOT IN ("SOFTWARE", "PHONE");
QUIT;
```

SAS Log Results

```
     PROC SQL;
       CREATE TABLE NOT_SO_HOT_PRODUCTS AS
         SELECT *
           FROM sql.PRODUCTS
             WHERE UPCASE(PRODTYPE) NOT IN ("SOFTWARE", "PHONE");
NOTE: Table NOT_SO_HOT_PRODUCTS created, with 3 rows and 5 columns.
     QUIT;
NOTE: PROCEDURE SQL used:
      real time           1.20 seconds
```

5.3 Populating Tables

After a table is created, it can then be populated with data. Unless the newly created table is defined as a subset of an existing table or its content is to remain static, one or more rows of data may eventually need to be added. The SQL standard provides the INSERT INTO statement as the vehicle for adding rows of data. The examples in this section look at how to add data in all the columns in a row as well as how to add data in only some of the columns in a row.

5.3.1 Adding Data to All the Columns in a Row

You populate tables with data by using an INSERT INTO statement. In fact, the INSERT INTO statement really doesn't insert rows of data at all. It simply adds each row to the end of the table. Three parameters are specified with an INSERT INTO statement: the name of the table, the names of the columns in which values are inserted, and the values themselves. Data values are inserted into a table with a VALUES clause. Suppose you want to insert (or add) a single row of data to the CUSTOMERS table and the row consists of three columns (Customer Number, Customer Name, and Home City).

SQL Code

```
PROC SQL;
  INSERT INTO CUSTOMERS (CUSTNUM, CUSTNAME, CUSTCITY)
     VALUES (702, 'Mission Valley Computing', 'San Diego');
QUIT;
```

The SAS log displays the following message noting that one row was inserted into the CUSTOMERS table.

SAS Log Results

```
PROC SQL;
  INSERT INTO CUSTOMERS
        (CUSTNUM, CUSTNAME, CUSTCITY)
         VALUES (702, 'Mission Valley Computing', 'San Diego');
NOTE: 1 row was inserted into CUSTOMERS.
  QUIT;
NOTE: PROCEDURE SQL used:
      real time           0.54 seconds
```

The inserted row of data from the previous INSERT INTO statement is added to the end of the CUSTOMERS table.

	Customer Number	Customer Name	Customer's Home City
1	101	La Mesa Computer Land	La Mesa
2	201	Vista Tech Center	Vista
3	301	Coronado Internet Zone	Coronado
4	401	La Jolla Computing	La Jolla
5	501	Alpine Technical Center	Alpine
6	601	Oceanside Computer Land	Oceanside
7	701	San Diego Byte Store	San Diego
8	801	Jamul Hardware & Software	Jamul
9	901	Del Mar Tech Center	Del Mar
10	1001	Lakeside Software Center	Lakeside
11	1101	Bonsall Network Store	Bonsall
12	1201	Rancho Santa Fe Tech	Rancho Santa Fe
13	1301	Spring Valley Byte Center	Spring Valley
14	1401	Poway Central	Poway
15	1501	Valley Center Tech Center	Valley Center
16	1601	Fairbanks Tech USA	Fairbanks Ranch
17	1701	Blossom Valley Tech	Blossom Valley
18	1801	Chula Vista Networks	
19	702	Mission Valley Computing	San Diego

Entering a new row into a table containing an index will automatically add the value to the index (for more information on indexes, see Chapter 6, "Modifying and Updating Tables and Indexes"). The following example illustrates adding three rows of data using the VALUES clause.

SQL Code

```
PROC SQL;
  INSERT INTO CUSTOMERS
            (CUSTNUM, CUSTNAME, CUSTCITY)
    VALUES (402, 'La Jolla Tech Center', 'La Jolla')
    VALUES (502, 'Alpine Byte Center',   'Alpine')
    VALUES (1702,'Rancho San Diego Tech','Rancho San Diego');

  SELECT *
    FROM CUSTOMERS
      ORDER BY CUSTNUM;
QUIT;
```

The SAS log shows that the three rows of data were inserted into the CUSTOMERS table.

SAS Log Results

```
     PROC SQL;
       INSERT INTO CUSTOMERS
                 (CUSTNUM, CUSTNAME, CUSTCITY)
         VALUES (402, 'La Jolla Tech Center', 'La Jolla')
         VALUES (502, 'Alpine Byte Center',   'Alpine')
         VALUES (1701,'Rancho San Diego Tech','Rancho San Diego');
NOTE: 3 rows were inserted into WORK.CUSTOMERS.

       SELECT *
         FROM CUSTOMERS
           ORDER BY CUSTNUM;
     QUIT;
NOTE: PROCEDURE SQL used:
      real time           1.03 seconds
```

The new rows are displayed in ascending order by CUSTNUM.

```
  Customer
    Number  Customer Name                Customer's Home City
  ----------------------------------------------------------
       101  La Mesa Computer Land        La Mesa
       201  Vista Tech Center            Vista
       301  Coronado Internet Zone       Coronado
       401  La Jolla Computing           La Jolla
       402  La Jolla Tech Center         La Jolla
       501  Alpine Technical Center      Alpine
       502  Alpine Byte Center           Alpine
       601  Oceanside Computer Land      Oceanside
       701  San Diego Byte Store         San Diego
       702  Mission Valley Computing     San Diego
       801  Jamul Hardware & Software    Jamul
       901  Del Mar Tech Center          Del Mar
      1001  Lakeside Software Center     Lakeside
      1101  Bonsall Network Store        Bonsall
      1201  Rancho Santa Fe Tech         Rancho Santa Fe
      1301  Spring Valley Byte Center    Spring Valley
```

(continued on next page)

```
1401  Poway Central                Poway
1501  Valley Center Tech Center    Valley Center
1601  Fairbanks Tech USA           Fairbanks Ranch
1701  Blossom Valley Tech          Blossom Valley
1702  Rancho San Diego Tech        Rancho San Diego
1801  Chula Vista Networks
```

5.3.2 Adding Data to Some of the Columns in a Row

It is not all that uncommon when adding rows of data to a table, to have one or more columns with an unassigned value. When this happens SQL must be able to handle adding the rows to the table as if all the values were present. But how does SQL handle values that are not specified? You will see in the following example that SQL assigns missing values to columns that do not have a value specified. As before, three parameters are specified with the INSERT INTO statement: the name of the table, the names of the columns in which values are inserted, and the values themselves. Suppose you had to add two rows of incomplete data to the CUSTOMERS table where two of three columns were specified (Customer Number and Customer Name).

SQL Code

```
PROC SQL;
  INSERT INTO CUSTOMERS
            (CUSTNUM, CUSTNAME)
     VALUES (102, 'La Mesa Byte & Floppy')
     VALUES (902, 'Del Mar Technology Center');

  SELECT *
    FROM CUSTOMERS
      ORDER BY CUSTNUM;
QUIT;
```

The SAS log shows the two rows of data added to the CUSTOMERS table.

SAS Log Results

```
     PROC SQL;
       INSERT INTO CUSTOMERS
                 (CUSTNUM, CUSTNAME)
         VALUES (102, 'La Mesa Byte & Floppy')
         VALUES (902, 'Del Mar Technology Center');
NOTE: 2 rows were inserted into WORK.CUSTOMERS.
       SELECT *
         FROM CUSTOMERS
           ORDER BY CUSTNUM;
     QUIT;
NOTE: PROCEDURE SQL used:
      real time           0.00 seconds
```

The new rows are displayed in ascending order by CUSTNUM with missing values assigned to the character column CUSTCITY.

```
  Customer
    Number  Customer Name                Customer's Home City
  ----------------------------------------------------------
       101  La Mesa Computer Land        La Mesa
       102  La Mesa Byte & Floppy
       201  Vista Tech Center            Vista
       301  Coronado Internet Zone       Coronado
       401  La Jolla Computing           La Jolla
       402  La Jolla Tech Center         La Jolla
       501  Alpine Technical Center      Alpine
       502  Alpine Byte Center           Alpine
       601  Oceanside Computer Land      Oceanside
       701  San Diego Byte Store         San Diego
       702  Mission Valley Computing     San Diego
       801  Jamul Hardware & Software    Jamul
       901  Del Mar Tech Center          Del Mar
       902  Del Mar Technology Center
      1001  Lakeside Software Center     Lakeside
      1101  Bonsall Network Store        Bonsall
```

(continued on next page)

```
    1201  Rancho Santa Fe Tech         Rancho Santa Fe
    1301  Spring Valley Byte Center    Spring Valley
    1401  Poway Central                Poway
    1501  Valley Center Tech Center    Valley Center
    1601  Fairbanks Tech USA           Fairbanks Ranch
    1701  Rancho San Diego Tech        Rancho San Diego
    1701  Blossom Valley Tech          Blossom Valley
    1801  Chula Vista Networks
```

In the previous example, missing values were assigned to the character column CUSTCITY. Suppose you want to add two rows of partial data to the PRODUCTS table where four of the five columns are specified (Product Number, Product Name, Product Type, and Product Cost) and the missing value for each row is the numeric column MANUNUM.

SQL Code

```
PROC SQL;
  INSERT INTO PRODUCTS
            (PRODNUM, PRODNAME, PRODTYPE, PRODCOST)
   VALUES(6002,'Security Software','Software',375.00)
   VALUES(1701,'Travel Laptop SE', 'Laptop',  4200.00);

  SELECT *
    FROM PRODUCTS
      ORDER BY PRODNUM;
QUIT;
```

The SAS log displays the two rows of data added to the PRODUCTS table.

SAS Log Results

```
     PROC SQL;
       INSERT INTO PRODUCTS
                  (PRODNUM, PRODNAME, PRODTYPE, PRODCOST)
        VALUES(6002,'Security Software','Software',375.00)
        VALUES(1701,'Travel Laptop SE', 'Laptop',  4200.00);
NOTE: 2 rows were inserted into WORK.PRODUCTS.

       SELECT *
         FROM PRODUCTS
           ORDER BY PRODNUM;
     QUIT;
NOTE: PROCEDURE SQL used:
      real time           0.75 seconds
```

The new rows are displayed in ascending order by PRODNUM with missing values assigned to the numeric column MANUNUM.

Product Number	Product Name	Manufacturer Number	Product Type	Product Cost
1110	Dream Machine	111	Workstation	$3,200.00
1200	Business Machine	120	Workstation	$3,300.00
1700	Travel Laptop	170	Laptop	$3,400.00
1701	**Travel Laptop SE**	**.**	**Laptop**	**$4,200.00**
2101	Analog Cell Phone	210	Phone	$35.00
2102	Digital Cell Phone	210	Phone	$175.00
2200	Office Phone	220	Phone	$130.00
5001	Spreadsheet Software	500	Software	$299.00
5002	Database Software	500	Software	$399.00
5003	Wordprocessor Software	500	Software	$299.00
5004	Graphics Software	500	Software	$299.00
6002	**Security Software**	**.**	**Software**	**$375.00**

5.3.3 Adding Data with a SELECT Query

You can also add data to a table using a SELECT query with an INSERT INTO statement. A query expression essentially executes an enclosed query by first creating a temporary table and then inserting the contents of the temporary table into the target table being populated. In the process of populating the target table, any columns omitted from the column list are automatically assigned to missing values.

In the next example, a SELECT query is used to add four rows of data from the SOFTWARE_PRODUCTS table into the PRODUCTS table. The designated query controls the insertion of data into the target PRODUCTS table using a WHERE clause.

SQL Code

```
PROC SQL;
  INSERT INTO PRODUCTS
             (PRODNUM, PRODNAME, PRODTYPE, PRODCOST)

  SELECT PRODNUM, PRODNAME, PRODTYPE, PRODCOST
    FROM SOFTWARE_PRODUCTS
      WHERE PRODTYPE IN ('Software');
QUIT;
```

The SAS log displays the results of the four rows of data added to the PRODUCTS table.

SAS Log Results

```
   PROC SQL;
     INSERT INTO PRODUCTS
                (PRODNUM, PRODNAME, PRODTYPE, PRODCOST)

     SELECT PRODNUM, PRODNAME, PRODTYPE, PRODCOST
       FROM SOFTWARE_PRODUCTS
         WHERE PRODTYPE IN ('Software');
NOTE: 4 rows were inserted into WORK.PRODUCTS.

   QUIT;
NOTE: PROCEDURE SQL used:
      real time           0.04 seconds
      cpu time            0.01 seconds
```

5.4 Integrity Constraints

The reliability of databases and the data within them is essential to every organization. Decision-making activities depend on the correctness and accuracy of any and all data contained in key applications, information systems, databases, decision support and query tools, as well as other critical systems. Even the slightest hint of unreliable data can affect decision-making capabilities, accuracy of reports, and, in those worst case scenarios, loss of user confidence in the database environment itself.

Because data should be correct and free of problems, an integral part of every database environment is a set of rules that the data should adhere to. These rules, often referred to as database-enforced constraints, are applied to the database table structure itself and determine the type and content of data that is permitted in columns and tables.

By implementing database-enforced integrity constraints, you can dramatically reduce data-related problems and additional programming work in applications. Instead of coding complex data checks and validations in individual application programs, you can build database-enforced constraints into the database itself. This work can eliminate the propagation of column duplication, invalid and missing values, lost linkages, and other data-related problems.

5.4.1 Defining Integrity Constraints

You define integrity constraints by specifying column definitions and constraints at the time a table is created with the CREATE TABLE statement or by adding, changing, or removing a table's column definitions with the ALTER TABLE statement. The rows in a table are then validated against the defined integrity constraints.

5.4.2 Types of Integrity Constraints

The first type of integrity constraint is referred to as a column and table constraint. This type of constraint essentially establishes rules that are attached to a specific table or column. The type of constraint is generally specified through one or two clauses with their distinct values as follows.

Column and Table Constraints

- NOT NULL
- UNIQUE
- CHECK

5.4.3 Preventing Null Values with a NOT NULL Constraint

A null value is essentially a missing or unknown value in the data. When unchecked, null values can often propagate themselves throughout a database. When a NULL appears in a mathematical equation, the returned result is also a null or missing value. When a NULL is used in a comparison or a logical expression, the returned result is unknown. The occurrence of null values presents problems during search, joins, and index operations. The ability to prevent the propagation of null values in a column with a NOT NULL constraint is a powerful feature of the SQL procedure. This constraint should be used as a first line of defense against potential problems resulting from the presence of null values and the interaction of queries processing data.

Using the CREATE TABLE or ALTER TABLE statement, you can apply a NOT NULL constraint to any column where missing, unknown, or inappropriate values appear in the data. Suppose you need to avoid the propagation of missing values in the CUSTCITY (Customer's Home City) column in the CUSTOMER_CITY table. By specifying the NOT NULL constraint for the CUSTCITY column in the CREATE TABLE statement, you prevent the propagation of null values in a table.

SQL Code

```
PROC SQL;
  CREATE TABLE CUSTOMER_CITY
     (CUSTNUM NUM,
      CUSTCITY CHAR(20) NOT NULL);
QUIT;
```

Once the CUSTOMER_CITY table is created and the NOT NULL constraint is defined for the CUSTCITY column, only non-missing data for the CUSTCITY column can be entered. Using the INSERT INTO statement with a VALUES clause, you can populate the CUSTOMER_CITY table while adhering to the assigned NOT NULL integrity constraint.

SQL Code

```
PROC SQL;
  INSERT INTO CUSTOMER_CITY
    VALUES(101,`La Mesa Computer Land')
    VALUES(1301,`Spring Valley Byte Center');
QUIT;
```

The SAS log shows the two rows of data satisfying the NOT NULL constraint and the rows successfully being added to the CUSTOMER_CITY table.

SAS Log Results

```
   PROC SQL;
     INSERT INTO CUSTOMER_CITY
       VALUES(101,'La Mesa Computer Land')
       VALUES(1301,'Spring Valley Byte Center');
NOTE: 2 rows were inserted into WORK.CUSTOMER_CITY.

   QUIT;
NOTE: PROCEDURE SQL used:
      real time          0.22 seconds
      cpu time           0.02 seconds
```

When you define a NOT NULL constraint and then attempt to populate a table with one or more missing data values, the rows will be rejected and the table restored to its original state. Essentially the insert fails because the NOT NULL constraint prevents any missing values for a defined column from populating a table. In the next example, several rows of data with a defined NOT NULL constraint are prevented from being populated in the CUSTOMER_CITY table because one row contains a missing CUSTCITY value.

SQL Code

```
PROC SQL;
  INSERT INTO CUSTOMER_CITY
    VALUES(101,'La Mesa Computer Land')
    VALUES(1301,'Spring Valley Byte Center')
    VALUES(1801,'');
QUIT;
```

The SAS log shows that the NOT NULL constraint has prevented the three rows of data from being populated in the CUSTOMER_CITY table. The violation caused an error message that resulted in the failure of the add/update operation. The UNDO_POLICY = REQUIRED option reverses all adds/updates that have been performed to the point of the error. This prevents errors or partial data from being propagated in the database table. The following SAS log results illustrate the error condition that caused the add/update operation to fail.

SAS Log Results

```
   PROC SQL;
     INSERT INTO CUSTOMER_CITY
       VALUES(101,'La Mesa Computer Land')
       VALUES(1301,'Spring Valley Byte Center')
       VALUES(1801,'');
ERROR: Add/Update failed for data set WORK.CUSTOMER_CITY because data
value(s) do not comply with integrity constraint _NM0001_.
NOTE: This insert failed while attempting to add data from VALUES
clause 3 to the data set.
NOTE: Deleting the successful inserts before error noted above to
restore table to a consistent state.
   QUIT;
NOTE: The SAS System stopped processing this step because of errors.
NOTE: PROCEDURE SQL used:
      real time           0.02 seconds
      cpu time            0.00 seconds
```

A NOT NULL constraint can also be applied to a column in an existing table containing data with an ALTER TABLE statement. To successfully impose a NOT NULL constraint you should not have missing or null values in the column the constraint is being defined for. This means that the presence of one or more null values in an existing table's column will prevent the NOT NULL constraint from being created.

Suppose the CUSTOMERS table contains one or more missing values in the CUSTCITY column. If you tried to add a NOT NULL constraint, it would be rejected. You can successfully apply the NOT NULL constraint only when missing values are reclassified or recoded.

SQL Code

```
PROC SQL;
  ALTER TABLE CUSTOMERS
    ADD CONSTRAINT NOT_NULL_CUSTCITY NOT NULL(CUSTCITY);
QUIT;
```

The SAS log shows the NOT NULL constraint cannot be defined in an existing table when a column's data contains one or more missing values. The violation produces an error message that results in the rejection of the constraint.

SAS Log Results

```
   PROC SQL;
     ALTER TABLE CUSTOMERS
       ADD CONSTRAINT NOT_NULL_CUSTCITY NOT NULL(CUSTCITY);
ERROR: Integrity constraint NOT_NULL_CUSTCITY was rejected because 1
observations failed the constraint.
   QUIT;
NOTE: The SAS System stopped processing this step because of errors.
NOTE: PROCEDURE SQL used:
      real time           0.01 seconds
      cpu time            0.00 seconds
```

5.4.4 Enforcing Unique Values with a UNIQUE Constraint

A UNIQUE constraint prevents duplicate values from propagating in a table. If you use a CREATE TABLE statement, you can apply a UNIQUE constraint to any column where duplicate data is not desired. Suppose you want to avoid the propagation of duplicate values in the CUSTNUM (Customer Number) column in a new table called CUSTOMER_CITY. By specifying the UNIQUE constraint for the CUSTNUM column with the CREATE TABLE statement, you prevent duplicate values from populating the table.

SQL Code

```
PROC SQL;
  CREATE TABLE CUSTOMER_CITY
     (CUSTNUM NUM UNIQUE,
      CUSTCITY CHAR(20));
QUIT;
```

When you define a UNIQUE constraint and attempt to populate a table with duplicate data values, the rows will be rejected and the table restored to its original state prior to the add operation taking place. Essentially the insert fails because the UNIQUE constraint prevents any duplicate values for a defined column from populating a table. In the next example, several rows of data with a defined UNIQUE constraint are prevented from being populated in the CUSTOMER_CITY table because one row contains a duplicate CUSTNUM value.

SQL Code

```
PROC SQL;
  INSERT INTO CUSTOMER_CITY
    VALUES(101,'La Mesa Computer Land')
    VALUES(1301,'Spring Valley Byte Center')
    VALUES(1301,'Chula Vista Networks');
QUIT;
```

The SAS log shows the UNIQUE constraint prevented the three rows of data from being populated in the CUSTOMER_CITY table. The violation caused an error message that resulted in the failure of the add/update operation.

SAS Log Results

```
  PROC SQL;
    INSERT INTO CUSTOMER_CITY
      VALUES(101,'La Mesa Computer Land')
      VALUES(1301,'Spring Valley Byte Center')
      VALUES(1301,'Chula Vista Networks');
ERROR: Add/Update failed for data set WORK.CUSTOMER_CITY because data
value(s) do not comply with integrity constraint _UN0001_.
NOTE: This insert failed while attempting to add data from VALUES
clause 3 to the data set.
NOTE: Deleting the successful inserts before error noted above to
restore table to a consistent state.
  QUIT;
NOTE: The SAS System stopped processing this step because of errors.
NOTE: PROCEDURE SQL used:
      real time           1.12 seconds
      cpu time            0.06 seconds
```

5.4.5 Validating Column Values with a CHECK Constraint

A CHECK constraint validates data values against a list of values, minimum and maximum values, as well as a range of values before populating a table. Using either a CREATE TABLE or ALTER TABLE statement, you can apply a CHECK constraint to any column that requires data validation to be performed. In the next example, suppose you want to validate data values in the PRODTYPE (Product Type) column in the PRODUCTS table. When you specify a CHECK constraint against the PRODTYPE column using the ALTER TABLE statement, product type values will first need to match the list of defined values or the rows will be rejected.

SQL Code

```
PROC SQL;
  ALTER TABLE PRODUCTS
    ADD CONSTRAINT CHECK_PRODUCT_TYPE
      CHECK (PRODTYPE IN ('Laptop',
                          'Phone',
                          'Software',
                          'Workstation'));
QUIT;
```

With a CHECK constraint defined, each row must meet the validation rules that are specified for the column before the table is populated. If any row does not pass the validation checks based on the established validation rules, the add/update operation fails and the table is automatically restored to its original state prior to the operation taking place. In the next example, three rows of data are validated against the defined CHECK constraint established for the PRODTYPE column.

SQL Code

```
PROC SQL;
  INSERT INTO PRODUCTS
    VALUES(5005,'Internet Software',500,'Software',99.)
    VALUES(1701,'Elite Laptop',170,'Laptop',3900.)
    VALUES(2103,'Digital Cell Phone',210,'Fone',199.);
QUIT;
```

The SAS log displays the results after attempting to add the three rows of data. Because one row violates the CHECK constraint with a value of "Fone", the rows are not added to the PRODUCTS table. The violation produced an error message that resulted in the failure of the add/update operation.

SAS Log Results

```
   PROC SQL;
     INSERT INTO PRODUCTS
       VALUES(5005,'Internet Software',500,'Software',99.)
       VALUES(1701,'Elite Laptop',170,'Laptop',3900.)
       VALUES(2103,'Digital Cell Phone',210,'Fone',199.);
ERROR: Add/Update failed for data set WORK.PRODUCTS because data
value(s) do not comply with integrity constraint CHECK_PRODUCT_TYPE.
NOTE: This insert failed while attempting to add data from VALUES
clause 3 to the data set.
NOTE: Deleting the successful inserts before error noted above to
restore table to a consistent state.
   QUIT;
NOTE: The SAS System stopped processing this step because of errors.
NOTE: PROCEDURE SQL used:
      real time           0.09 seconds
      cpu time            0.02 seconds
```

5.4.6 Referential Integrity Constraints

The second type of constraint that is available in the SQL procedure is referred to as a referential integrity constraint. Enforced through primary and foreign keys between two or more tables, referential integrity constraints are built into a database environment to prevent data integrity issues from occurring. The specific types of referential integrity constraints and constraint action clauses are used to enforce update and delete operations and consist of the following:

Referential Integrity Constraints

- Primary key
- Foreign key

Referential Integrity Constraint Action Clauses

- RESTRICT (Default)
- SET NULL
- CASCADE

The action clauses are discussed in the section, "Establishing a Foreign Key."

5.4.7 Establishing a Primary Key

A primary key consists of one or more columns with a unique value that is used to identify individual rows in a table. Depending on the nature of the columns used, a single column may be all that is necessary to identify specific rows. In other cases, two or more columns may be needed to adequately identify a row in a referenced table. Suppose you needed to uniquely identify specific rows in the MANUFACTURERS table. By establishing the Manufacturer Number (MANUNUM) as the unique identifier for rows, a key is established. The next example specifies the ALTER TABLE statement to create a primary key using MANUNUM in the MANUFACTURERS table.

SQL Code

```
PROC SQL;
  ALTER TABLE MANUFACTURERS
    ADD CONSTRAINT PRIM_KEY PRIMARY KEY (MANUNUM);
QUIT;
```

The SAS log shows that the MANUFACTURERS table has been modified successfully after creating a primary key using the MANUNUM column.

SAS Log Results

```
   PROC SQL;
     ALTER TABLE MANUFACTURERS
       ADD CONSTRAINT PRIM_KEY PRIMARY KEY (MANUNUM);
NOTE: Table WORK.MANUFACTURERS has been modified, with 4 columns.
   QUIT;
NOTE: PROCEDURE SQL used:
      real time           0.07 seconds
      cpu time             0.01 seconds
```

Suppose you also needed to uniquely identify specific rows in the PRODUCTS table. By specifying PRODNUM (Product Number) as the primary key, the next example specifies the ALTER TABLE statement to create the unique identifier for rows in the table.

SQL Code

```
PROC SQL;
  ALTER TABLE PRODUCTS
    ADD CONSTRAINT PRIM_PRODUCT_KEY PRIMARY KEY (PRODNUM);
QUIT;
```

The SAS log shows that the PRODUCTS table has been modified successfully after establishing a primary key using the PRODNUM column.

SAS Log Results

```
   PROC SQL;
     ALTER TABLE PRODUCTS
       ADD CONSTRAINT PRIM_PRODUCT_KEY PRIMARY KEY (PRODNUM);
NOTE: Table WORK.PRODUCTS has been modified, with 5 columns.
   QUIT;
NOTE: PROCEDURE SQL used:
      real time           0.03 seconds
      cpu time            0.01 seconds
```

5.4.8 Establishing a Foreign Key

A foreign key consists of one or more columns in a table that references or relates to values in another table. The column(s) used as a foreign key must match the column(s) in the table that is referenced. The purpose of a foreign key is to ensure that rows of data in one table exist in another table thereby preventing the possibility of lost or missing linkages between tables. The enforcement of referential integrity rules has a positive and direct effect on data reliability issues.

Suppose you wanted to ensure that data values in the INVENTORY table have corresponding and matching data values in the PRODUCTS table. By establishing PRODNUM (Product Number) as a foreign key in the INVENTORY table you ensure a strong level of data integrity between the two tables. This essentially verifies that key data in the INVENTORY table exists in the PRODUCTS table. In the next example a foreign key is created using the PRODNUM column in the INVENTORY table by specifying the ALTER TABLE statement.

SQL Code

```
PROC SQL;
 ALTER TABLE INVENTORY
  ADD CONSTRAINT FOREIGN_PRODUCT_KEY FOREIGN KEY (PRODNUM)
   REFERENCES PRODUCTS
    ON DELETE RESTRICT
    ON UPDATE RESTRICT;
QUIT;
```

The SAS log displays the successful creation of the PRODNUM column as a foreign key in the INVENTORY table. By specifying the default values ON DELETE RESTRICT and ON UPDATE RESTRICT clauses, you restrict the ability to change the values of primary key data when matching values are found in the foreign key. The execution of any SQL statement that could violate these referential integrity rules is prevented during SQL processing.

SAS Log Results

```
   PROC SQL;
    ALTER TABLE INVENTORY
     ADD CONSTRAINT FOREIGN_PRODUCT_KEY FOREIGN KEY (PRODNUM)
      REFERENCES PRODUCTS
       ON DELETE RESTRICT
       ON UPDATE RESTRICT;
NOTE: Table WORK.INVENTORY has been modified, with 5 columns.
   QUIT;
NOTE: PROCEDURE SQL used:
      real time           0.01 seconds
      cpu time            0.01 seconds
```

Suppose a product of a particular manufacturer is no longer available and has been taken off the market. To handle this type of situation, data values in the INVENTORY table should be set to missing once the product is deleted from the PRODUCTS table. The next example establishes a foreign key using the PRODNUM column in the INVENTORY table and sets values to null with the ON DELETE clause.

SQL Code

```
PROC SQL;
 ALTER TABLE INVENTORY
  ADD CONSTRAINT FOREIGN_MISSING_PRODUCT_KEY FOREIGN KEY
(PRODNUM)
   REFERENCES PRODUCTS
    ON DELETE SET NULL;
QUIT;
```

The SAS log displays the successful creation of the PRODNUM column as a foreign key in the INVENTORY table as well as the effect of the ON DELETE SET NULL clause. Specifying this clause will change foreign key values to missing or null for all rows matching values found in the primary key. The execution of any SQL statement that could violate these referential integrity rules is prevented during SQL processing.

SAS Log Results

```
   PROC SQL;
    ALTER TABLE INVENTORY
     ADD CONSTRAINT FOREIGN_MISSING_PRODUCT_KEY FOREIGN KEY (PRODNUM)
      REFERENCES PRODUCTS
       ON DELETE SET NULL;
NOTE: Table WORK.INVENTORY has been modified, with 5 columns.
   QUIT;
NOTE: PROCEDURE SQL used:
      real time           0.02 seconds
      cpu time            0.02 seconds
```

Suppose you want to ensure that changes to key values in the PRODUCTS table automatically flow over or cascade through to rows in the INVENTORY table. This is accomplished by first creating PRODNUM (Product Number) as a foreign key in the INVENTORY table using the ADD CONSTRAINT clause and referencing the PRODUCTS table. You then specify the ON UPDATE CASCADE clause to enable any changes made to the PRODUCTS table to be automatically cascaded through to the INVENTORY table. This ensures that changes to the product number values in the PRODUCTS table automatically occur in the INVENTORY table as well.

SQL Code

```
PROC SQL;
 ALTER TABLE INVENTORY
  ADD CONSTRAINT FOREIGN_PRODUCT_KEY FOREIGN KEY (PRODNUM)
   REFERENCES PRODUCTS
    ON UPDATE CASCADE
    ON DELETE RESTRICT   /* DEFAULT VALUE */;
QUIT;
```

The SAS log displays the successful creation of the PRODNUM column as a foreign key in the INVENTORY table. When the ON UPDATE and ON DELETE clauses are specified, the execution of any SQL statement that could violate referential integrity rules is strictly prohibited.

SAS Log Results

```
   PROC SQL;
    ALTER TABLE INVENTORY
     ADD CONSTRAINT FOREIGN_PRODUCT_KEY FOREIGN KEY (PRODNUM)
      REFERENCES PRODUCTS
       ON UPDATE CASCADE
      ON DELETE RESTRICT   /* DEFAULT VALUE */;
NOTE: Table WORK.INVENTORY has been modified, with 5 columns.
   QUIT;
NOTE: PROCEDURE SQL used:
      real time           0.06 seconds
      cpu time            0.01 seconds
```

5.4.8.1 Constraints and Change Control

To preserve change control the SAS System prohibits changes or modifications to a table containing a defined referential integrity constraint. When you attempt to delete, rename, or replace a table containing a referential integrity constraint, an error message is generated and processing is stopped. The next example illustrates a table copy operation that is performed against a table containing a referential integrity constraint that generates an error message and stops processing.

SAS Log Results

```
   PROC COPY IN=SQLBOOK OUT=WORK;
     SELECT INVENTORY;
   RUN;

NOTE: Copying SQLBOOK.INVENTORY to WORK.INVENTORY (memtype=DATA).
ERROR: A rename/delete/replace attempt is not allowed for a data set
involved in a referential integrity constraint. WORK.INVENTORY.DATA
ERROR: File WORK.INVENTORY.DATA has not been saved because copy could
not be completed.
NOTE: Statements not processed because of errors noted above.
NOTE: PROCEDURE COPY used:
      real time             0.44 seconds
      cpu time              0.02 seconds

NOTE: The SAS System stopped processing this step because of errors.
```

5.4.9 Displaying Integrity Constraints

Using the DESCRIBE TABLE statement, the SQL procedure displays integrity constraints along with the table description on the SAS log. The ability to capture this type of information assists with the documentation process by describing the names and types of integrity constraints as well as the contributing columns they reference.

SQL Code

```
PROC SQL;
  DESCRIBE TABLE MANUFACTURERS;
QUIT;
```

The SAS log shows the SQL statements that were used to create the MANUFACTURERS table as well as an alphabetical list of integrity constraints that have been defined.

SAS Log Results

```
   PROC SQL;
     DESCRIBE TABLE MANUFACTURERS;
NOTE: SQL table WORK.MANUFACTURERS was created like:

create table WORK.MANUFACTURERS( bufsize=4096 )
  (
   manunum num label='Manufacturer Number',
   manuname char(25) label='Manufacturer Name',
   manucity char(20) label='Manufacturer City',
   manustat char(2) label='Manufacturer State'
  );
create unique index manunum on WORK.MANUFACTURERS(manunum);

                 -----Alphabetic List of Integrity Constraints-----

                            Integrity
                       #    Constraint     Type            Variables
                       ___________________________________________

                       1    PRIM_KEY       Primary Key     manunum
   QUIT;
NOTE: PROCEDURE SQL used:
      real time           0.19 seconds
      cpu time            0.01 seconds
```

5.5 Deleting Rows in a Table

In the world of data management, the ability to delete unwanted rows of data from a table is as important as being able to populate a table with data. In fact, data management activities would be severely hampered without the ability to delete rows of data. The DELETE statement and an optional WHERE clause can remove one or more unwanted rows from a table, depending on what is specified in the WHERE clause.

5.5.1 Deleting a Single Row in a Table

The DELETE statement can be specified to remove a single row of data by constructing an explicit WHERE clause on a unique value. The construction of a WHERE clause to satisfy this form of row deletion may require a complex logic construct. So be sure to test the expression thoroughly before applying it to the table to determine whether it performs as expected. The following example illustrates the removal of a single customer in the CUSTOMERS table by specifying the customer's name (CUSTNAME) in the WHERE clause.

SQL Code

```
PROC SQL;
  DELETE FROM CUSTOMERS2
    WHERE UPCASE(CUSTNAME) = "LAUGHLER";
QUIT;
```

SAS Log Results

```
      PROC SQL;
        DELETE FROM CUSTOMERS2
          WHERE UPCASE(CUSTNAME) = "LAUGHLER";
NOTE: 1 row was deleted from WORK.CUSTOMERS2.
      QUIT;
NOTE: PROCEDURE SQL used:
      real time           0.37 seconds
```

5.5.2 Deleting More Than One Row in a Table

Frequently, a row deletion affects more than a single row in a table. In these cases a WHERE clause references a value occurring multiple times. The following example illustrates the removal of a single customer in the PRODUCTS table by specifying the product type (PRODTYPE) in the WHERE clause.

SQL Code

```
PROC SQL;
  DELETE FROM PRODUCTS
    WHERE UPCASE(PRODTYPE) = `PHONE';
QUIT;
```

SAS Log Results

```
     PROC SQL;
       DELETE FROM PRODUCTS
         WHERE UPCASE(PRODTYPE) = `PHONE';
NOTE: 3 rows were deleted from WORK.PRODUCTS.
     QUIT;
NOTE: PROCEDURE SQL used:
      real time           0.05 seconds
```

5.5.3 Deleting All Rows in a Table

SQL provides a simple way to delete all rows in a table. The following example shows that all rows in the CUSTOMERS table can be removed when the WHERE clause is omitted. Use care when using this form of the DELETE statement because every row in the table is automatically deleted.

SQL Code

```
PROC SQL;
  DELETE FROM CUSTOMERS;
QUIT;
```

SAS Log Results

```
     PROC SQL;
       DELETE FROM CUSTOMERS;
NOTE: 28 rows were deleted from WORK.CUSTOMERS.
     QUIT;
NOTE: PROCEDURE SQL used:
      real time           0.00 seconds
```

5.6 Deleting Tables

The SQL standard permits one or more unwanted tables to be removed (or deleted) from a database (SAS library). During large program processes, temporary tables in the WORK library are frequently created. The creation and build-up of these tables can negatively affect memory and storage performance areas, causing potential problems due to insufficient resources. It is important from a database management perspective to be able to delete any unwanted tables to avoid these types of resource problems. Here are a few guidelines to keep in mind.

Before a table can be deleted, complete ownership of the table (that is, exclusive access to the table) should be verified. Although some SQL implementations require a table to be empty in order to delete it, the SAS implementation permits a table to be deleted with or without any rows of data in it. After a table is deleted, any references to that table are no longer recognized and will result in a syntax error. Additionally, any references to a deleted table in a view will also result in an error (see Chapter 8, "Working with Views"). Also, any indexes associated with a deleted table are automatically dropped (see Chapter 6, "Modifying and Updating Tables and Indexes").

5.6.1 Deleting a Single Table

Deleting a table from the database environment is not the same as making a table empty. Although an empty table contains no data, it still possesses a structure; a deleted table contains no data or related structure. Essentially a deleted table does not exist because the table including its data and structure are physically removed forever. Deleting a single table from a database environment requires a single table name to be referenced in a DROP TABLE statement. In the next example, a single table called HOT_PRODUCTS located in the WORK library is physically removed using a DROP TABLE statement as follows.

SQL Code

```
PROC SQL;
  DROP TABLE HOT_PRODUCTS;
QUIT;
```

SAS Log Results

```
     PROC SQL;
       DROP TABLE HOT_PRODUCTS;
NOTE: Table WORK.HOT_PRODUCTS has been dropped.
     QUIT;
NOTE: PROCEDURE SQL used:
      real time           0.38 seconds
```

5.6.2 Deleting Multiple Tables

The SQL standard also permits more than one table to be specified in a single DROP TABLE statement. The next example and corresponding log shows two tables (HOT_PRODUCTS and NOT_SO_HOT_PRODUCTS) being deleted from the WORK library.

SQL Code

```
PROC SQL;
  DROP TABLE HOT_PRODUCTS, NOT_SO_HOT_PRODUCTS;
QUIT;
```

SAS Log Results

```
     PROC SQL;
       DROP TABLE HOT_PRODUCTS, NOT_SO_HOT_PRODUCTS;
NOTE: Table WORK.HOT_PRODUCTS has been dropped.
NOTE: Table WORK.NOT_SO_HOT_PRODUCTS has been dropped.
     QUIT;
NOTE: PROCEDURE SQL used:
      real time           0.00 seconds
```

5.6.3 Deleting Tables Containing Integrity Constraints

As previously discussed in this chapter, to ensure a high-level of data integrity in a database environment the SQL standard permits the creation of one or more integrity constraints to be imposed on a table. Under the SQL standard, a table containing one or more constraints cannot be deleted without first dropping the defined constraints. This behavior further safeguards and prevents the occurrence of unanticipated surprises such as the accidental deletion of primary or supporting tables.

In the next example, the SAS log shows that an error is produced when an attempt to drop a table containing an ON DELETE RESTRICT referential integrity constraint is performed. The referential integrity constraint caused the DROP TABLE statement to fail resulting in the INVENTORY table not being deleted.

SAS Log

```
   PROC SQL;
     DROP TABLE INVENTORY;
ERROR: A rename/delete/replace attempt is not allowed for a data set
involved in a referential integrity constraint. WORK.INVENTORY.DATA
WARNING: Table WORK.INVENTORY has not been dropped.
   QUIT;
NOTE: The SAS System stopped processing this step because of errors.
NOTE: PROCEDURE SQL used:
      real time           0.00 seconds
      cpu time            0.00 seconds
```

To enable the deletion of a table containing one or more integrity constraints, you must specify an SQL statement such as the ALTER TABLE statement and DROP COLUMN or DROP CONSTRAINT clauses. Once a table's integrity constraints are removed, the table can then be deleted.

In the following SAS log, the FOREIGN_PRODUCT_KEY constraint is removed from the INVENTORY table using the DROP CONSTRAINT clause. With the constraint removed, the INVENTORY table is then deleted with the DROP TABLE statement.

SAS Log Results

```
   PROC SQL;
     ALTER TABLE INVENTORY
       DROP CONSTRAINT FOREIGN_PRODUCT_KEY;
NOTE: Integrity constraint FOREIGN_PRODUCT_KEY deleted.
NOTE: Table WORK.INVENTORY has been modified, with 5 columns.
   QUIT;
NOTE: PROCEDURE SQL used:
      real time           0.01 seconds
      cpu time            0.01 seconds



   PROC SQL;
     DROP TABLE INVENTORY;
NOTE: Table WORK.INVENTORY has been dropped.
   QUIT;
NOTE: PROCEDURE SQL used:
      real time           0.34 seconds
      cpu time            0.02 seconds
```

5.7 Summary

1. Creating a table using a column-definition list is similar to defining a table's structure with the LENGTH statement in the DATA step (see section 5.2.1).

2. Using the LIKE clause copies the column names and attributes from the existing table structure to the new table structure (see section 5.2.2).

3. Deriving a table from an existing table stores the results in a new table instead of displaying them as SAS output (see section 5.2.3).

4. In populating a table, three parameters are specified with an INSERT INTO statement: the name of the table, the names of the columns in which values are inserted, and the values themselves (see section 5.3.1).

5. Database-enforced constraints can be applied to a database table structure to enforce the type and content of data that is permitted (see section 5.4).

6. The DELETE statement combined with a WHERE clause selectively removes one or more rows of data from a table (see section 5.5.1).

7. The SQL standard permits one or more unwanted tables to be removed from a database (SAS library) (see section 5.6.2).

Chapter 6

Modifying and Updating Tables and Indexes

6.1 Introduction

After a table is defined and populated with data, a column as well as its structure may need modifying. The SQL standard provides Data Definition Language (DDL) statements to permit changes to a table's structure and its data. In this chapter, you will see examples that add and delete columns, modify column attributes, add and delete indexes, rename tables, and update values in rows of data.

6.2 Modifying Tables

An important element in PROC SQL is its Data Definition Language (DDL) capabilities. From creating and deleting tables (see Chapter 5, "Creating, Populating, and Deleting Tables") and indexes to altering table structures and columns, the DDL provides PROC SQL programmers with a way to change (or redefine) the definition of one or more existing tables. The ALTER TABLE statement permits columns to be added, modified, or dropped in a table with the ADD, MODIFY, or DROP clauses. When a table's columns or attributes are modified, the table's structural dynamics also change. The following sections examine the various ways tables can be modified in the SQL procedure.

6.2.1 Adding New Columns

As requirements and needs change, a database's initial design may require one or more new columns to be added. Before any new columns can be added, complete ownership of the table must be granted. When you have exclusive access, each new column that you add is automatically added at the end of the table's descriptor record. This means that the ALTER TABLE statement's ADD clause modifies the table without reading or writing data.

Suppose you were given a new requirement to improve your ability to track the status of inventory levels. It is determined that your organization can achieve this new capability by adding a new column to the INVENTORY table. The ADD clause is used in the ALTER TABLE statement to define the new column, INVENTORY_STATUS, and its attributes. The new column's purpose is to identify the following inventory status values: "In-Stock", "Out-of-Stock", and "Back Ordered".

SQL Code

```
PROC SQL;
  ALTER TABLE INVENTORY
    ADD inventory_status char(12);
QUIT;
```

Once the new column is added, the SAS log indicates that 6 columns exist in the INVENTORY table.

SAS Log Results

```
     PROC SQL;
       ALTER TABLE INVENTORY
         ADD inventory_status char(12);
NOTE: Table WORK.INVENTORY has been modified, with 6
columns.
     QUIT;
NOTE: PROCEDURE SQL used:
      real time           0.44 seconds
```

The output shows the INVENTORY_STATUS column added at the end of the INVENTORY table.

Results

```
                             The CONTENTS Procedure

           -----Alphabetic List of Variables and Attributes-----

#  Variable          Type   Len  Pos  Format     Informat  Label
------------------------------------------------------------------------------------
4  invencst          Num      6   22  DOLLAR10.2           Inventory Cost
2  invenqty          Num      3   19                       Inventory Quantity
6  inventory_status  Char    12    4
5  manunum           Num      3   28                       Manufacturer Number
3  orddate           Num      4    0  MMDDYY10.  MMDDYY10. Date Inventory Last Ordered
1  prodnum           Num      3   16                       Product Number
```

(continued on next page)

-----Variables Ordered by Position-----

#	Variable	Type	Len	Pos	Format	Informat	Label
1	prodnum	Num	3	16			Product Number
2	invenqty	Num	3	19			Inventory Quantity
3	orddate	Num	4	0	MMDDYY10.	MMDDYY10.	Date Inventory Last Ordered
4	invencst	Num	6	22	DOLLAR10.2		Inventory Cost
5	manunum	Num	3	28			Manufacturer Number
6	INVENTORY_STATUS	Char	12	4			

6.2.2 Controlling the Position of Columns in a Table

Column position is not normally important in relational database processing. But, there are times when a particular column order is desired, for example when SELECT * (select all) syntax is specified. To add one or more columns in a designated order, the SQL standard provides a couple of choices. You can:

1. Create a new table with the columns in the desired order and load the data into the new table,

2. Create a view that puts the columns in the desired order and then access the view in lieu of the table (see Chapter 8, “Working with Views,” for a detailed explanation).

Suppose you had to add the INVENTORY_STATUS column so it is inserted between the ORDDATE and INVENCST columns and not just added as the last column in the table. The following example shows how this can be done. As before, we begin by adding the INVENTORY_STATUS column to the INVENTORY table. Then, we create a new table called INVENTORY_COPY and load the data from INVENTORY in the following column order: PRODNUM, INVENQTY, ORDDATE, INVENTORY_STATUS, INVENCST, and MANUNUM.

SQL Code

```
PROC SQL;
  ALTER TABLE INVENTORY
    ADD INVENTORY_STATUS CHAR(12);
  CREATE TABLE INVENTORY_COPY AS
    SELECT PRODNUM, INVENQTY, ORDDATE, INVENTORY_STATUS,
           INVENCST, MANUNUM
      FROM INVENTORY;
QUIT;
PROC CONTENTS DATA=INVENTORY_COPY;
RUN;
```

The PROC CONTENTS output below shows the positioning of the columns in the new INVENTORY_COPY table including the new INVENTORY_STATUS column that was added.

Results

The CONTENTS Procedure

-----Variables Ordered by Position-----

#	Variable	Type	Len	Pos	Format	Informat	Label
1	prodnum	Num	3	16			Product Number
2	invenqty	Num	3	19			Inventory Quantity
3	orddate	Num	4	0	MMDDYY10.	MMDDYY10.	Date Inventory Last Ordered
4	INVENTORY_STATUS	Char	12	4			
5	invencst	Num	6	22	DOLLAR10.2		Inventory Cost
6	manunum	Num	3	28			Manufacturer Number

Another way of controlling a table's column order is to create a view or virtual table (for more information on views, see Chapter 8, "Working with Views"), from an existing table by specifying the desired column order. Using a CREATE VIEW statement and a SELECT query you can construct a new view so that the columns appear in a desired order. Essentially the view contains no data, just the PROC SQL query's instructions that were used to create it. The biggest advantage of creating a view to reorder the columns defined in a table is that a view not only avoids the creation of a physical table, but hides sensitive data from unauthorized viewing. In the next example, a new view called

INVENTORY_VIEW is created from the INVENTORY table with selected columns appearing in a specific order.

SQL Code

```
PROC SQL;
  CREATE VIEW INVENTORY_VIEW AS
    SELECT PRODNUM, INVENQTY, INVENTORY_STATUS
      FROM INVENTORY;
QUIT;
```

The PROC CONTENTS output below shows the positioning of the columns in the new view including the new INVENTORY_STATUS column that was added earlier.

Results

```
                         The CONTENTS Procedure

      -----Alphabetic List of Variables and Attributes-----

 #    Variable              Type    Len    Pos    Label
 ____________________________________________________________________

 2    invenqty              Num       3      8    Inventory Quantity
 3    inventory_status      Char     12     11
 1    prodnum               Num       3      0    Product Number
```

6.2.3 Changing a Column's Length

Column definitions (length, informat, format, and label) can be modified with the MODIFY clause in the ALTER TABLE statement. PROC SQL enables a character or numeric column to have its length changed. In the next example, suppose you had to reduce the length of the character column MANUCITY in the MANUFACTURERS table from 20 bytes to a length of 15 bytes to conserve space. The CHAR column-definition is used in the MODIFY clause in the ALTER TABLE statement to redefine the length of the column.

SQL Code

```
PROC SQL;
  ALTER TABLE MANUFACTURERS
    MODIFY MANUCITY CHAR(15);
QUIT;
```

SAS Log Results

```
     PROC SQL;
       ALTER TABLE MANUFACTURERS
         MODIFY MANUCITY CHAR(15);
NOTE: Table WORK.MANUFACTURERS has been modified, with 4 columns.
     QUIT;
NOTE: PROCEDURE SQL used:
      real time           0.50 seconds
```

The PROC CONTENTS output below illustrates the changed column length made to the MANUCITY column in the MANUFACTURERS table.

Results

```
                           The CONTENTS Procedure

  Data Set Name: WORK.MANUFACTURERS                 Observations:          6
  Member Type:   DATA                               Variables:             4
  Engine:        V8                                 Indexes:               0
  Created:       14:21 Tuesday, November 9, 1999    Observation Length:    45

       -----Alphabetic List of Variables and Attributes-----

          #    Variable    Type    Len    Pos    Label
          ___________________________________________________________
          3    manucity    Char     15     25    Manufacturer City
          2    manuname    Char     25      0    Manufacturer Name
          1    manunum     Num       3     42    Manufacturer Number
          4    manustat    Char      2     40    Manufacturer State
```

The column length can also be changed using the PROC SQL LENGTH= option in the SELECT clause of the CREATE TABLE statement. This construct avoids your having to use the ALTER TABLE statement, as illustrated in the previous example, as well as using a DATA step. The next example shows the LENGTH= option to reduce the length of the MANUCITY column from 20 bytes to 15 bytes.

SQL Code

```
PROC SQL;
  CREATE TABLE MANUFACTURERS_MODIFIED AS
    SELECT MANUNUM, MANUNAME, MANUCITY LENGTH=15, MANUSTAT
      FROM MANUFACTURERS;
QUIT;
```

SAS Log Results

```
  PROC SQL;
    CREATE TABLE MANUFACTURERS_MODIFIED AS
      SELECT MANUNUM, MANUNAME, MANUCITY LENGTH=15, MANUSTAT
        FROM MANUFACTURERS;
NOTE: Table WORK.MANUFACTURERS_MODIFIED created, with 6 rows and 4 columns.

  QUIT;
NOTE: PROCEDURE SQL used (Total process time):
      real time           0.12 seconds
      cpu time            0.01 seconds
```

A column that is initially defined as numeric can also have its length changed in PROC SQL. The SQL procedure ignores a field width in these situations and defines all numeric columns with a maximum width of 8 bytes. The reason is that numeric columns are always defined with the maximum precision allowed by the SAS System. To override this limitation, it is recommended that you use a LENGTH= option in the SELECT clause of the CREATE TABLE statement or the LENGTH statement in a DATA step to assign (or reassign) any numeric column lengths to the desired size. You can also improve query results by assigning indexes only to those columns that have many unique values or that you use regularly in joins.

In the next example, the numeric column MANUNUM has its length changed (or redefined) from 3 bytes to 4 bytes using the LENGTH= option in the SELECT clause of the CREATE TABLE statement.

Note: Recursive references in the target table can create data integrity problems. For this reason you should refrain from specifying the same table name in the CREATE TABLE statement as specified in the FROM clause.

SQL Code

```
PROC SQL;
  CREATE TABLE MANUFACTURERS_MODIFIED AS
    SELECT MANUNUM LENGTH=4, MANUNAME, MANUCITY, MANUSTAT
      FROM MANUFACTURERS;
QUIT;
```

The PROC CONTENTS output illustrates the changed column length assigned to the numeric MANUNUM column in the MANUFACTURERS_MODIFIED table.

Results

```
                        The CONTENTS Procedure


           Alphabetic List of Variables and Attributes

     #     Variable     Type     Len     Label

     3     manucity     Char      15     Manufacturer City
     2     manuname     Char      25     Manufacturer Name
     1     manunum      Num        4     Manufacturer Number
     4     manustat     Char       2     Manufacturer State
```

In the next example, the numeric column MANUNUM has its length changed (or redefined) from 3 bytes to 4 bytes using the LENGTH statement in a DATA step. To avoid truncation or data problems, you should verify that a column having a shorter length can handle existing data. Because PROC SQL does not produce any notes or warnings if numeric values are truncated, you are required to know your data.

DATA Step Code

```
DATA MANUFACTURERS;
  LENGTH MANUNUM 4.;
  SET MANUFACTURERS;
RUN;
```

SAS Log Results

```
     DATA MANUFACTURERS;
       LENGTH MANUNUM 4.;
       SET MANUFACTURERS;
     RUN;

NOTE: There were 6 observations read from the dataset
WORK.MANUFACTURERS.
NOTE: The data set WORK.MANUFACTURERS has 6 observations and 4
variables.
NOTE: DATA statement used:
      real time           0.44 seconds
```

The PROC CONTENTS output below illustrates the changed column length assigned to the numeric MANUNUM column in the MANUFACTURERS table.

Results

```
                          The CONTENTS Procedure

        -----Alphabetic List of Variables and Attributes-----

   #    Variable     Type    Len     Pos     Label
   ___________________________________________________________
   3    manucity     Char     15      29     Manufacturer City
   2    manuname     Char     25       4     Manufacturer Name
   1    manunum      Num       4       0     Manufacturer Number
   4    manustat     Char      2      44     Manufacturer State
```

6.2.4 Changing a Column's Format

You can permanently change a column's format with the MODIFY clause of the ALTER TABLE statement—and not just for the duration of the step. Suppose you had to increase the size of the DOLLAR*w.d* format from DOLLAR9.2 to DOLLAR12.2 to allow larger product cost (PRODCOST) values in the PRODUCTS table to print properly.

SQL Code

```
PROC SQL;
  ALTER TABLE PRODUCTS
    MODIFY PRODCOST FORMAT=DOLLAR12.2;
QUIT;
```

SAS Log Results

```
     PROC SQL;
       ALTER TABLE PRODUCTS
         MODIFY PRODCOST FORMAT=DOLLAR12.2;
NOTE: Table WORK.PRODUCTS has been modified, with 5 columns.
     QUIT;
NOTE: PROCEDURE SQL used:
      real time            0.33 seconds
```

6.2.5 Changing a Column's Label

You can modify a column's label information with the ALTER TABLE statement MODIFY clause. Because the label information is part of the descriptor record, changes to this value have no impact on the data itself. Suppose you had to change the label corresponding to the product cost (PRODCOST) column in the PRODUCTS table so when printed it displayed "Retail Product Cost".

SQL Code

```
PROC SQL;
  ALTER TABLE PRODUCTS
    MODIFY PRODCOST LABEL="Retail Product Cost";
QUIT;
```

SAS Log Results

```
     PROC SQL;
        ALTER TABLE PRODUCTS
        MODIFY PRODCOST LABEL="Retail Product Cost";
NOTE: Table WORK.PRODUCTS has been modified, with 5 columns.
     QUIT;
NOTE: PROCEDURE SQL used:
      real time           0.00 seconds
```

6.2.6 Renaming a Column

The SQL procedure does provide an ANSI approach to renaming columns in a table. By specifying the SELECT clause in the CREATE TABLE statement, you can rename columns, although it can be tedious if a large number of columns exist in the table. The next example illustrates a SELECT clause in a CREATE TABLE statement being used to rename the ITEM column to ITEM_PURCHASED in the PURCHASES table. As the example below illustrates you should refrain from specifying the same table name in the CREATE TABLE statement as specified in the FROM clause. Recursive references to the target table can cause data integrity problems.

SQL Code

```
PROC SQL;
  CREATE TABLE PURCHASES AS
    SELECT CUSTNUM, ITEM AS ITEM_PURCHASED, UNITS, UNITCOST
      FROM PURCHASES;
QUIT;
```

SAS Log Results

```
   PROC SQL;
     CREATE TABLE PURCHASES AS
       SELECT CUSTNUM, ITEM AS ITEM_PURCHASED, UNITS, UNITCOST
         FROM PURCHASES;
WARNING: This CREATE TABLE statement recursively references the
target table. A consequence of this is a possible data integrity
problem.
NOTE: Table WORK.PURCHASES created, with 7 rows and 4 columns.

   QUIT;
NOTE: PROCEDURE SQL used (Total process time):
      real time           0.41 seconds
      cpu time            0.02 seconds
```

An alternative approach to renaming columns in a table consists of using the **RENAME=** SAS data set option in a SELECT statement's FROM clause. Suppose you needed to rename ITEM in the PURCHASES table to ITEM_PURCHASED. In the next example, the RENAME= SAS data set option can be specified in one of two ways, as illustrated below. Either approach is syntactically correct.

SQL Code

```
PROC SQL;
  SELECT *
    FROM PURCHASES (RENAME=ITEM=ITEM_PURCHASED);
QUIT;
```

< or >

```
PROC SQL;
  SELECT *
    FROM PURCHASES (RENAME=(ITEM=ITEM_PURCHASED));
QUIT;
```

SAS Log Results

```
      PROC SQL;
        SELECT *
          FROM PURCHASES(RENAME=ITEM=ITEM_PURCHASED);
      QUIT;
NOTE: PROCEDURE SQL used:
       real time           0.31 seconds
       cpu time            0.02 seconds
```

6.2.7 Renaming a Table

The SQL procedure does not provide a standard ANSI approach to renaming a table in a SAS library. Consequently, the DATASETS procedure is the recommended method to accomplish this relatively simple task. Suppose you had to rename the PRODUCTS table in the WORK library to MANUFACTURED_PRODUCTS.

SAS Code

```
PROC DATASETS LIBRARY=WORK;
  CHANGE PRODUCTS = MANUFACTURED_PRODUCTS;
RUN;
```

SAS Log Results

```
PROC DATASETS LIBRARY=work;
                                   Directory

           Libref          WORK
           Engine          V9
           Physical Name   D:\SAS Version 9.1\SAS Temporary Files\_TD1704
           File Name       D:\SAS Version 9.1\SAS Temporary Files\_TD1704


                                      Member    File
        #  Name                       Type      Size  Last Modified

        1  CUSTOMERS                  DATA      5120  16Aug04:23:37:30
        2  CUSTOMERS2                 DATA      5120  16Aug04:23:37:30
        3  INVENTORY                  DATA      5120  16Aug04:23:39:22
        4  INVOICE                    DATA      5120  16Aug04:23:37:32
        5  MANUFACTURERS              DATA      5120  17Aug04:00:07:40
        6  PRODUCTS                   DATA     17408  17Aug04:00:10:38
        7  PURCHASES                  DATA      5120  17Aug04:00:17:12
      CHANGE PRODUCTS = MANUFACTURED_PRODUCTS;
    RUN;

NOTE: Changing the name WORK.PRODUCTS to WORK.MANUFACTURED_PRODUCTS (memtype=DATA).
```

An assortment of novel approaches has been used to rename tables. One approach, shown below, uses the CREATE TABLE statement with the SELECT query to create a new table with the desired table name followed by the DROP TABLE statement to delete the old table. You should be aware, however, that this is not an efficient method to rename a table.

SQL Code

```
PROC SQL;
  CREATE TABLE MANUFACTURED_PRODUCTS AS
    SELECT *
      FROM PRODUCTS;
  DROP TABLE PRODUCTS;
QUIT;
```

6.3 Indexes

An index consists of one or more columns used to uniquely identify each row within a table. Operating as a SAS object containing the values in one or more columns in a table, an index is composed of one or more columns and may be defined as numeric, character, or a combination of both.

There is no rule that says a table has to have an index, but they can often make information retrieval more efficient and considerably faster.

For example, if you know a specific part number and its location from a list of thousands, then you can look up the part and find its manufacturer, cost, and location far more efficiently than if you did not know this information.

6.3.1 Defining Indexes

When defining an index, you should first understand the purpose the index is to serve. The most important thing to keep in mind about indexes is that they should be created only when they are absolutely needed. Too many, or unnecessary, indexes use up computer resources. An index also takes up space and has to be updated any time a DELETE, INSERT, or UPDATE is performed on rows in a table. For this reason, care should be used when deciding when and what indexes to create.

To help determine when indexes are necessary, consider existing data as well as the way the base table(s) will be used. You also need to know what queries will be used and how they will access columns of data. If an index is used to specify some order within a table, such as manufacturer number or product number in the PRODUCTS table, you should fully assess what the impact of that index will be.

Sometimes the column(s) making up an index is obvious, and other times it is not. When determining whether an index provides any value, some very important rules should be kept in mind. An index should permit the greatest flexibility so every column in a table can be accessed and displayed. You can also improve query results by assigning indexes only to those columns that have many unique values or that you use regularly in joins.

When an index is specified for one or more tables, a join process may actually occur faster. The PROC SQL processor may use an index when certain conditions permit its use. Here are a few things to keep in mind before creating an index:

- If the table is small, sequential processing may be just as fast, or faster, than processing with an index
- If the page count as displayed in the CONTENTS procedure is less than 3 pages, avoid creating or using an index
- Do not create more indexes than you absolutely need
- If the data subset for the index is not small, sequential access may be more efficient than using the index
- If the percentage of matches is approximately 15% or less then an index should be used
- The costs associated with an index can outweigh its performance value – an index is updated each time when rows in a table are added, deleted, or modified.

Two types of indexes can be defined and used in PROC SQL: *simple* and *composite.* When a simple index is created, it references only a single column. In contrast, a composite index references two or more columns in a table.

6.3.2 Creating a Simple Index

A simple index is specifically defined for one column in a table and must be the **same** name as the column. Suppose you had to create an index consisting of product type (PRODTYPE) in the PRODUCTS table. Once created, the index becomes a separate object located in the SAS library.

SQL Code

```
PROC SQL;            ❶             ❷           ❸
  CREATE INDEX PRODTYPE ON PRODUCTS(PRODTYPE);
QUIT;
```

SAS Log Results

```
     PROC SQL;            ❶             ❷           ❸
       CREATE INDEX PRODTYPE ON PRODUCTS(PRODTYPE);
NOTE: Simple index PRODTYPE has been defined.
     QUIT;
NOTE: PROCEDURE SQL used:
      real time           0.37 seconds
```

❶ The simple index is assigned a name of PRODTYPE, which must be the same as the column name.

❷ The simple index is defined on the PRODUCTS table.

❸ The PRODTYPE column in the PRODUCTS table is designated as the column to be used by the index.

6.3.3 Creating a Composite Index

A composite index is specifically defined for two or more columns in a table and must have a different name from the columns. Suppose you had to create an index consisting of manufacturer number (MANUNUM) and product type (PRODTYPE) located in the PRODUCTS table. You should be aware that only one composite index is allowed per set of columns, but more than one composite index is allowed. The composite index, as with the simple index, becomes a separate object located in the SAS library.

SQL Code

```
PROC SQL;
  CREATE INDEX    ❶             ❷             ❸
         MANUNUM_PRODTYPE ON PRODUCTS(MANUNUM,PRODTYPE);
QUIT;
```

SAS Log Results

```
     PROC SQL;
       CREATE INDEX       ❶              ❷               ❸
              MANUNUM_PRODTYPE ON PRODUCTS(MANUNUM,PRODTYPE);
NOTE: Composite index MANUNUM_PRODTYPE has been defined.
     QUIT;
NOTE: PROCEDURE SQL used:
      real time           0.00 seconds
```

❶ The composite index is assigned a name of MANUNUM_PRODTYPE, which is used to represent the MANUNUM and PRODTYPE column names.

❷ The composite index is defined on the PRODUCTS table.

❸ The MANUNUM and PRODTYPE columns in the PRODUCTS table are designated as the columns to be used by the index.

6.3.4 Preventing Duplicate Values in an Index

The UNIQUE keyword prevents the entry of a duplicate value in an index. You should use this keyword with care because there may be times when more than one occurrence of a data value in a table is necessary. When multiple occurrences of the same value appear in a table, the UNIQUE keyword is rejected and the index is not created for that particular column.

6.3.5 Modifying Columns Containing Indexes

Altering the attributes of a column that contains an associated index (simple or composite) does NOT prohibit the values in the altered column from using the index. But, if a column that contains an index is dropped, then the index is also dropped. Accordingly, when a column is dropped, any data in that index is also lost.

6.3.6 Deleting (Dropping) Indexes

When one or more indexes are no longer needed, the DROP INDEX statement can be used to remove them. Suppose you determine that you no longer need the composite index MANUNUM_PRODTYPE (created earlier) because processing requirements have changed. The next example illustrates a single composite index being deleted from the SAS library.

SQL Code

```
PROC SQL;
  DROP INDEX MANUNUM_PRODTYPE
    FROM PRODUCTS;
QUIT;
```

SAS Log Results

```
      PROC SQL;
      DROP INDEX MANUNUM_PRODTYPE
         FROM PRODUCTS;
NOTE: Index MANUNUM_PRODTYPE has been dropped.
      QUIT;
NOTE: PROCEDURE SQL used:
      real time           0.00 seconds
```

According to the ANSI SQL standard, two or more indexes can also be deleted in a DROP INDEX statement. The next example illustrates the MANUNUM and PRODTYPE indexes being deleted from the SAS library in a single DROP INDEX statement.

SQL Code

```
PROC SQL;
  DROP INDEX MANUNUM, PRODTYPE
    FROM PRODUCTS;
QUIT;
```

SAS Log Results

```
  PROC SQL;
    DROP INDEX MANUNUM, PRODTYPE
      FROM PRODUCTS;
NOTE: Index MANUNUM has been dropped.
NOTE: Index PRODTYPE has been dropped.
  QUIT;
NOTE: PROCEDURE SQL used (Total process time):
      real time           0.00 seconds
      cpu time            0.01 seconds
```

6.4 Updating Data in a Table

Once a table is populated with data, you may need to update values in one or more of its rows. Column values in existing rows in a table can be updated with the UPDATE statement. The key to successful row updates is the creation of a well-constructed SET clause and WHERE expression. If the WHERE expression is not constructed correctly, the possibility of an update error is great.

Suppose all laptops in the PRODUCTS table have just been discounted by 20 percent and the new price is to take effect immediately. The update would compute the discounted product cost for "Laptop" computers only. For example, the discounted price for a laptop computer would be reduced to $2,720.00 from $3,400.00.

SQL Code

```
PROC SQL;
  UPDATE PRODUCTS
    SET PRODCOST = PRODCOST - (PRODCOST * 0.2)
      WHERE UPCASE(PRODTYPE) = 'LAPTOP';

  SELECT *
    FROM PRODUCTS;
QUIT;
```

SAS Log Results

```
      PROC SQL;
        UPDATE PRODUCTS
          SET PRODCOST = PRODCOST - (PRODCOST * 0.2)
            WHERE UPCASE(PRODTYPE) = 'LAPTOP';
NOTE: 1 row was updated in WORK.PRODUCTS.
        SELECT *
          FROM PRODUCTS;
      QUIT;
NOTE: PROCEDURE SQL used:
       real time            0.00 seconds
```

Results

Product Number	Product Name	Manufacturer Number	Product Type	Retail Product Cost
1110	Dream Machine	111	Workstation	$3,200.00
1200	Business Machine	120	Workstation	$3,300.00
1700	**Travel Laptop**	**170**	**Laptop**	**$2,720.00**
2101	Analog Cell Phone	210	Phone	$35.00
2102	Digital Cell Phone	210	Phone	$175.00
2200	Office Phone	220	Phone	$130.00
5001	Spreadsheet Software	500	Software	$299.00
5002	Database Software	500	Software	$399.00
5003	Wordprocessor Software	500	Software	$299.00
5004	Graphics Software	500	Software	$299.00

6.5 Summary

1. Data Definition Language (DDL) statements provide programmers with a way to redefine the definition of one or more existing tables (see section 6.2).
2. As one or more new columns are added to a table, each is automatically added at the end of a table's descriptor record (see section 6.2.1).

3. To add one or more columns in a designated order, the SQL standard provides a couple of choices to choose from (see section 6.2.2).
4. PROC SQL enables a character column (but not a numeric column) to have its length changed (see section 6.2.3).
5. A column's format and label information can be modified with a MODIFY clause (see sections 6.2.4 and 6.2.5).
6. The RENAME= SAS data set option must be used in a FROM clause to rename column names (see section 6.2.6).
7. The DATASETS procedure is the recommended way to rename tables (see section 6.2.7).
8. An index consists of one or more columns used to uniquely identify each row within a table (see section 6.3).
9. Column values in existing rows in a table can be modified with the UPDATE statement (see section 6.4).

Chapter 7

Coding Complex Queries

7.1 Introduction

In previous chapters, our discussion of queries was confined to a single table referenced with a SELECT statement. The real strength of the relational approach is the ability it gives you to construct queries that refer to several tables or even to other queries. These types of queries are referred to as *complex queries*. PROC SQL provides a way to construct complex queries by enabling you to join two or more tables, build queries that control other queries through a process known as nesting, and combine output as a single table from multiple queries.

7.2 Introducing Complex Queries

In the previous chapters, the queries could be classified as being relatively simple because references were always made against a single table. We now turn our attention to queries of a more complex nature that call on the full features of the SQL procedure. Four complex query constructs will be illustrated in this chapter.

Inner Joins	Up to 32 tables are referenced in a FROM and optional WHERE clause of a SELECT statement.
Outer Joins	A maximum of two tables are referenced in a FROM and ON clause of a SELECT statement.
Subqueries	A query is embedded (nested) in the WHERE clause of a main query.
Set Operations	A new results table is created from two separate queries.

7.3 Joins

Joining two or more tables of data is a powerful feature in the relational model. The SQL procedure enables you to join tables of information quickly and easily. Linking one piece of information with another piece of information is made possible when at least one column is common to each table. A maximum of 32 tables can be combined using conventional (inner) join techniques, as opposed to a maximum of two tables at a time using outer join techniques.

This chapter discusses a number of join topics including why joins are important, the differences between the various join techniques, the importance of the WHERE clause in creating joins, creating and using table aliases, joining three or more tables of data, outer (left, right, and full) joins, subqueries, and set operations. It is important to recognize that many of these techniques can be accomplished using DATA step programming techniques, but the simplicity and flexibility found in the SQL procedure makes it especially useful, if not indispensable, as a tool for the practitioner.

7.3.1 Why Joins Are Important

As relational database systems continue to grow in popularity, the need to access normalized data stored in separate tables becomes increasingly important. By relating matching values in key columns in one table with key columns in the other table(s), you can retrieve information as if the data were stored in one huge file. The results can provide new and exciting insights into possible data relationships.

7.3.2 Information Retrieval Based on Relationships

Being able to define relationships between multiple tables and retrieve information based on these relationships is a powerful feature of the relational model. A join of two or more tables provides a means of gathering and manipulating data in a single SELECT statement. You join two or more tables by specifying the table names in a SELECT statement. Joins are specified on a minimum of two tables at a time, where a column from each table is used for the purpose of connecting the two tables. Connecting columns should have *"like"* values and the same column attributes because the join's success is dependent on these values.

In a typical join, you name the relevant columns in the SELECT statement, you specify the tables to be joined in the FROM clause, and in the WHERE clause you specify the relationship you want revealed. That is, you describe the data subset that you want to produce. To be of use (and of a manageable size) your join needs a WHERE clause to constrain the results and ensure their utility and relevance.

Note: When you create a join without a WHERE clause, you are creating an internal, virtual table called a Cartesian product. This table can be extremely large because it represents all possible combinations of rows and columns in the joined tables.

7.3.3 Types of Complex Queries

The SQL procedure supports a great number of complex queries (sometimes referred to as join types). From inner joins to left, right, and full outer joins, this chapter provides a comprehensive look at the various forms of SELECT statements that can be used to perform multiple table management. Additional topics and examples include subqueries and set operations such as UNION, INTERSECT, and EXCEPT operations. The next table presents the various types of complex queries available in the SQL procedure.

Types of Complex Queries

Query Type	Description
Cartesian Product or Cross Join	This type of join creates a table representing all the combinations of rows and columns from two or more tables. It is represented by the absence of a WHERE clause.
Inner Joins	This type of join is referred to as a conventional type of join because it only retrieves rows with matching values from two or more tables (maximum of 32 tables).
Equijoin	A join with an equality condition (for example, equal sign "=") specified between columns in two or more tables.
Non-Equijoin	A join with an inequality condition (for example, NE, >, <) specified between columns in two or more tables.
Reflexive or Self Join	A join that combines a table with itself.
Outer Joins	A join that retrieves rows with matching values while preserving some or all of the unmatched rows from one or both tables.
Left Outer Join	A join that preserves unmatched rows from the left table.
Right Outer Join	A join that preserves unmatched rows from the right table.
Full Outer Join	A join that preserves unmatched rows from the left and right tables.

Query Type	Description
Subqueries	A query within another query – sometimes referred to as a nested query that retrieves rows from one table based on values in another table.
Simple Subquery	A self-contained and independent query within another query that returns single or multiple values from an inner query.
Correlated Subquery	An outer query that passes value(s) to an inner query that after execution passes the results back to the outer query.
Set Operations	These operators combine or concatenate query results vertically.
UNION	Combines all unique (nonduplicate) rows from both queries.
INTERSECT	Combines all matched rows from the first query with rows in the second query.
EXCEPT	Produces rows from the first query that do not appear in the second query.
OUTER UNION	Concatenates (appends) the results from both queries.

7.4 Cartesian Product Joins

As mentioned previously, the Cartesian product (or cross join) represents all possible combinations of rows and columns from the joined tables. To be exact, it represents the sum of the number of columns of the input tables plus the product of the number of rows of the input tables. Put another way, it represents each row from the first table matched with each possible row from the second table, and so on and so forth. For example, if you performed a joint operation on one table consisting of 100,000 rows and a second table of 10,000 rows, you would get a Cartesian product of 10 million rows.

Although the Cartesian product serves a very useful purpose in the relational model, it is essentially meaningless for a user to intentionally produce it as a final table. Besides being large, Cartesian products contain too much information and make it difficult, if not impossible, for the practitioner to select what is salient. It is only when you subset the Cartesian product using a WHERE clause that your data becomes quantifiable and manageable. For more information on Cartesian Product joins and examples illustrating the results of these joins, go to the Companion Web Site for this book.

7.5 Inner Joins

As was mentioned earlier, inner joins can handle a maximum of 32 tables at a time, and are the most recognized and widely used type of join. They are principally used to restrict rows where the specific search condition is not met. As a result, only rows satisfying the conditions specified in the WHERE clause are kept. This is in direct contrast with outer joins (discussed in a later section).

7.5.1 Equijoins

The most common form of inner join often referred to as an *equijoin* uses an equal sign "=" in the WHERE clause to indicate equality between the columns in two or more tables. Suppose you wanted to match products with their corresponding manufacturers so that all products from each manufacturer would be listed. An equijoin is performed to equate the manufacturer number from tables PRODUCTS and MANUFACTURERS.

SQL Code

```
PROC SQL;
  SELECT prodname, prodcost,
         manufacturers.manunum, manuname
            ➊           ➋
          FROM PRODUCTS, MANUFACTURERS
       WHERE products.manunum =          ➌
             manufacturers.manunum;
QUIT;
```

➊ The PRODUCTS table is the first table specified in the FROM clause.

➋ The MANUFACTURERS table is the second table specified in the FROM clause.

➌ The specification of an equal sign "=" in a WHERE clause between the columns in the tables indicates an equality type of join.

Results

```
                        The SAS System

                        Product  Manufacturer
Product Name               Cost  Number  Manufacturer Name
------------------------------------------------------------------
Dream Machine         $3,200.00     111  Cupid Computer
Business Machine      $3,300.00     120  Storage Devices Inc
Analog Cell Phone        $35.00     210  Global Comm Corp
Digital Cell Phone      $175.00     210  Global Comm Corp
Spreadsheet Software    $299.00     500  KPL Enterprises
Database Software       $399.00     500  KPL Enterprises
Wordprocessor Software  $299.00     500  KPL Enterprises
Graphics Software       $299.00     500  KPL Enterprises
```

The previous example can be further qualified by adding another condition in the WHERE clause. For example, suppose you wanted to display only those products from the manufacturer KPL Enterprises. The following join identifies all the products manufactured by KPL Enterprises as specified in the WHERE clause (all rows not meeting the condition of the WHERE clause are automatically excluded from the results of the join).

Note: This join assumes you know KPL Enterprises's unique manufacturer number.

SQL Code

```
PROC SQL;
  SELECT prodname, prodcost,
         manufacturers.manunum, manuname
    FROM PRODUCTS, MANUFACTURERS
      WHERE products.manunum =          ❶
            manufacturers.manunum    AND
            products.manunum = 500;
QUIT;
```

❶ The specification of the AND logical operator in the WHERE clause indicates that both conditions must be true in order to retrieve rows from both tables.

Results

```
                          The SAS System

                          Product  Manufacturer
Product Name                 Cost  Number   Manufacturer Name
-----------------------------------------------------------------
Spreadsheet Software      $299.00       500  KPL Enterprises
Database Software         $399.00       500  KPL Enterprises
Wordprocessor Software    $299.00       500  KPL Enterprises
Graphics Software         $299.00       500  KPL Enterprises
```

Let's extend our knowledge of equijoins a bit further by identifying how much money is tied up with products manufactured by KPL Enterprises. To accomplish this, you need to do two things. First, you need to sum the product cost (PRODCOST) column across all rows that match the WHERE clause condition. Because the objective of the equijoin is to compute a total amount for products manufactured by KPL Enterprises, you need to prevent duplicate rows from displaying in the result. To do so, specify the DISTINCT keyword.

SQL Code

```
PROC SQL;
  SELECT DISTINCT SUM(prodcost) AS Total_Cost     ❶
                  FORMAT=DOLLAR10.2,
         manufacturers.manunum
    FROM PRODUCTS, MANUFACTURERS
      WHERE products.manunum =
            manufacturers.manunum AND
            manufacturers.manuname = 'KPL Enterprises';


QUIT;
```

❶ The DISTINCT keyword prevents duplicate rows from appearing in the result.

Results

```
                The SAS System

    Total_Cost    Manufacturer Name
    ---------------------------------------
     $1,296.00    KPL Enterprises
```

7.5.2 Non-Equijoins

Another type of inner join is known as a non-equijoin. As you might guess from its name, a non-equijoin does not have an equal sign "=" specified in its WHERE clause. For example, suppose you want to display products manufactured by KPL Enterprises that cost more than $299.00. The use of the greater than ">"operator gives this type of join its name.

Note: When the SQL procedure optimizer is unable to optimize a join query by reducing the Cartesian product, a message is displayed in the SAS log indicating that the join requires performing one or more Cartesian product joins and cannot be optimized.

SQL Code

```
PROC SQL;
  SELECT prodname, prodtype, prodcost,
         manufacturers.manunum, manufacturers.manuname
    FROM PRODUCTS, MANUFACTURERS
      WHERE manufacturers.manunum = 500 AND
            prodtype = 'Software' AND
            prodcost > 299.00;    ❶
QUIT;
```

❶ The specification of the greater than ">"operator in the WHERE clause indicates a non-equijoin scenario.

SAS Log Results

```
   PROC SQL;
     SELECT prodname, prodtype, prodcost,
            manufacturers.manunum, manufacturers.manuname
       FROM PRODUCTS, MANUFACTURERS
         WHERE manufacturers.manunum = 500 AND
               prodtype = 'Software' AND
               prodcost > 299.00;
NOTE: The execution of this query involves performing one or more Cartesian
product joins that can not be optimized.
   QUIT;
NOTE: PROCEDURE SQL used:
      real time           0.01 seconds
      cpu time            0.01 seconds
```

Results

```
                              The SAS System

                   Product          Product  Manufacturer   Manufacturer
Product Name       Type                Cost        Number   Name
--------------------------------------------------------------------------
Database Software  Software         $399.00           500   KPL Enterprises
```

7.5.3 Reflexive or Self Joins

The final type of inner join is referred to as a reflexive join, or as it is sometimes called by practitioners a self join. As its name implies, a self join makes an internal copy of a table and joins the copy to itself. Essentially a join of this type joins one copy of a table to itself for the purpose of exploiting and illustrating comparisons between table values. For example, suppose you want to compare the prices of products side-by-side by product type with the less expensive product appearing first (in the first three columns of example result below).

SQL Code

```
PROC SQL;
  SELECT products.prodname, products.prodtype,
           products.prodcost,
        products_copy.prodname, products_copy.prodtype,
           products_copy.prodcost
            ❶          ❷
    FROM PRODUCTS, PRODUCTS PRODUCTS_COPY
      WHERE products.prodtype =                  ❸
              products_copy.prodtype AND
           products.prodcost <
              products_copy.prodcost;
QUIT;
```

❶ The PRODUCTS table is the primary table specified in the FROM clause.

❷ A copy of the PRODUCTS table called PRODUCTS_COPY is joined with the PRODUCTS table.

❸ The WHERE clause requests the same type of products to be compared side-by-side with the less expensive product appearing first.

Results

The SAS System

Product Name ❶	Product Type	Product Cost	Product Name ❷	Product Type	Product Cost
❸ Dream Machine	Workstation	$3,200.00	Business Machine	Workstation	$3,300.00
Analog Cell Phone	Phone	$35.00	Digital Cell Phone	Phone	$175.00
Analog Cell Phone	Phone	$35.00	Office Phone	Phone	$130.00
Office Phone	Phone	$130.00	Digital Cell Phone	Phone	$175.00
Spreadsheet Software	Software	$299.00	Database Software	Software	$399.00
Wordprocessor Software	Software	$299.00	Database Software	Software	$399.00
Graphics Software	Software	$299.00	Database Software	Software	$399.00

Looking at another example, suppose you want to find out the names and invoice amounts where, for each customer, you list the names and invoice amounts of each customer with larger invoice amounts. The next example illustrates a very useful application of a self join.

SQL Code

```
PROC SQL;
  SELECT invoice.custnum, invoice.invprice,
         invoice_copy.custnum, invoice_copy.invprice
             ❶             ❷
    FROM INVOICE, INVOICE INVOICE_COPY
      WHERE invoice.invprice <           ❸
                invoice_copy.invprice;
QUIT;
```

❶ The INVOICE table is the primary table specified in the FROM clause.

❷ A copy of the INVOICE table called INVOICE_COPY is joined with the INVOICE table.

❸ The WHERE clause produces names of customers with larger invoice amounts.

Results

The SAS System

Customer Number	Invoice Unit Price	Customer Number	Invoice Unit Price
201	$1,495.00	1301	$1,598.00
201	$1,495.00	501	$9,600.00
201	$1,495.00	401	$23,100.00
1301	$1,598.00	501	$9,600.00
1301	$1,598.00	401	$23,100.00
101	$245.00	201	$1,495.00
101	$245.00	1301	$1,598.00
101	$245.00	501	$9,600.00
101	$245.00	801	$798.00
101	$245.00	901	$396.00
101	$245.00	401	$23,100.00
501	$9,600.00	401	$23,100.00
801	$798.00	201	$1,495.00
801	$798.00	1301	$1,598.00

(continued on next page)

```
801         $798.00         501      $9,600.00
801         $798.00         401     $23,100.00
901         $396.00         201      $1,495.00
901         $396.00        1301      $1,598.00
901         $396.00         501      $9,600.00
901         $396.00         801        $798.00
901         $396.00         401     $23,100.00
```

7.5.4 Using Table Aliases in Joins

Every table in a SAS library must have a unique name to reference it. Table names must conform to valid SAS naming conventions having a maximum length of 32 characters and starting with a letter or underscore (see the *SAS Language Reference: Concepts* for further details).

To minimize the number of keystrokes when referencing the tables specified in a join query, you can assign an *alias* or temporary table name reference to each table. When assigned, these arbitrary aliases provide a short-cut method to the tables themselves and are in effect for the duration of the join query *but no longer*. In the next example, the table alias "P" is assigned to the PRODUCTS table and the alias "M" is assigned to the MANUFACTURERS table in the FROM clause. Table name references in the SELECT statement and WHERE clause are made easier as well.

SQL Code

```
PROC SQL;
  SELECT prodnum, prodname, prodtype, M.manunum
    FROM PRODUCTS  P, MANUFACTURERS  M              ❶
      WHERE P.manunum  = M.manunum AND
            M.manuname = 'KPL Enterprises';
QUIT;
```

❶ The assignment of the table alias "P" and the table alias "M" in the FROM clause provides a short-cut method of referencing the longer table names PRODUCTS and MANUFACTURERS.

Results

```
                          The SAS System

 Product                                               Manufacturer
  Number  Product Name                 Product Type          Number
 ------------------------------------------------------------------
    5001  Spreadsheet Software         Software                 500
    5002  Database Software            Software                 500
    5003  Wordprocessor Software       Software                 500
    5004  Graphics Software            Software                 500
```

7.5.5 Performing Computations in Joins

Join queries, as with simpler queries, can take full advantage of the power of the SQL procedure. Logical and arithmetic operators, predicates, and summary functions are all available for you to use. The join query is an essential component because stored information is not always available in the form we need.

PROC SQL provides the ability to perform basic arithmetic operations such as addition, subtraction, multiplication, and division with columns containing numeric values. Essentially, this enables any query to perform column addition, subtraction, multiplication, and division. Suppose you had to compute the sales tax of 7.75% for all manufactured products sold in the state of California. In the next example, the SELECT statement shows the California sales tax (using the product cost column and the fixed sales tax percentage) computation, assigns a column alias to the result column as well as a format and label to enhance the readability of the result.

SQL Code

```
PROC SQL;
  SELECT prodname, prodtype, prodcost,
         prodcost * .0775 AS SalesTax      ❶
         FORMAT=dollar10.2  LABEL='California Sales Tax'
    FROM PRODUCTS  P, MANUFACTURERS  M
      WHERE P.manunum = M.manunum AND
            M.manustat = 'CA';
QUIT;
```

❶ The ability to perform basic arithmetic operations in a SELECT statement as well as assign a column alias to the result is part of the SQL ANSI standard.

Results

```
                              The SAS System

                                                                  ❶
                                                     Product  California
Product Name                 Product Type               Cost   Sales Tax
------------------------------------------------------------------------
Business Machine             Workstation           $3,300.00     $255.75
Analog Cell Phone            Phone                    $35.00       $2.71
Digital Cell Phone           Phone                   $175.00      $13.56
Spreadsheet Software         Software                $299.00      $23.17
Database Software            Software                $399.00      $30.92
Wordprocessor Software       Software                $299.00      $23.17
Graphics Software            Software                $299.00      $23.17
```

7.5.6 Joins with Three Tables

Up to this point, our examples have been limited to two-table joins. But what if more information is needed than the two tables can provide? To extract the required information, access to a third table may be necessary. A join with three tables is a fairly simple extension of a two-table join.

As before, each joinable column must possess the same column attributes and contain the same type of information. Besides listing all required tables in the FROM clause, the WHERE clause would need to include any and all restrictions to subset only the rows desired. For example, suppose you want to display only those products along with their invoice quantity that appear in the INVOICE table for the manufacturer KPL Enterprises (manunum=500).

SQL Code

```
PROC SQL;
  SELECT P.prodname,
         P.prodcost,
         M.manuname,
         I.invqty
    FROM PRODUCTS  P,
         MANUFACTURERS  M,
         INVOICE  I
      WHERE P.manunum = M.manunum AND
            P.prodnum = I.prodnum AND
            M.manunum = 500;
QUIT;
```

Results

The SAS System

Product Name	Product Cost	Manufacturer Name	Invoice Quantity - Units Sold
Spreadsheet Software	$299.00	KPL Enterprises	5
Database Software	$399.00	KPL Enterprises	2

Let's examine the construction of the WHERE clause for this three-way join a bit further. The column containing the manufacturer number from the PRODUCTS, MANUFACTURERS, and INVOICE tables is joined using an AND logical operator in the WHERE clause. Additionally, the WHERE clause restricts the resulting table to only product invoices for manufacturer (manunum=500). In the next example, a three-way join lists the product names and costs, along with the customer who bought each product.

SQL Code

```
PROC SQL;
  SELECT P.prodname,
         P.prodcost,
         C.custname,
         I.invprice
    FROM PRODUCTS  P,
         INVOICE   I,
         CUSTOMERS C
      WHERE P.prodnum = I.prodnum AND
            I.custnum = C.custnum;
QUIT;
```

Results

```
                          The SAS System

                        Product                                 Invoice
Product Name               Cost  Customer Name                    Price
------------------------------------------------------------------------
Analog Cell Phone        $35.00  La Mesa Computer Land          $245.00
Spreadsheet Software    $299.00  Vista Tech Center            $1,495.00
Business Machine      $3,300.00  La Jolla Computing          $23,100.00
Dream Machine         $3,200.00  Alpine Technical Center      $9,600.00
Database Software       $399.00  Jamul Hardware & Software      $798.00
```

7.5.7 Joins with More Than Three Tables

Occasionally, information needs to be extracted from four, five, or more tables (up to a maximum of 32 tables). Joins of four or more tables can be constructed just like those accessing two or three tables. The only difference is the number of table references in the FROM clause and the level of complexity in the WHERE clause to restrict what rows are kept. Suppose you want to know, based on invoices, the number of products ordered before September 1, 2000. One way to find this information is to perform a join with four tables.

SQL Code

```
PROC SQL;
  SELECT sum(inventory.invenqty)
           AS Products_Ordered_Before_09012000
    FROM PRODUCTS,
         INVOICE,
         CUSTOMERS,
         INVENTORY
      WHERE inventory.orddate < mdy(09,01,00) AND
            products.prodnum  = invoice.prodnum AND
            invoice.custnum   = customers.custnum AND
            invoice.prodnum   = inventory.prodnum;
QUIT;
```

Results

```
                 The SAS System

                    Products_
              Ordered_Before_
                     09012000
                    _________
                            8
```

If you were wondering whether this result could have been derived another way, you would be correct. You could also determine, based on invoices, the number of products ordered before September 1, 2000, with the following two-way join code. As can be seen, there is often more than one way to construct a join to extract the information you want.

SQL Code

```
PROC SQL;
  SELECT sum(inventory.invenqty)
           AS Products_Ordered_Before_09012000
    FROM INVOICE   I,
         INVENTORY I2
      WHERE inventory.orddate < mdy(09,01,00) AND
            invoice.prodnum   = inventory.prodnum;
QUIT;
```

Results

```
                    The SAS System

                       Products_
                 Ordered_Before_
                        09012000
                       ________
                               8
```

To expand your understanding of joins with more than three tables, we will illustrate a four-table join. Suppose you want to know the products being purchased and who is purchasing them. The next example shows a four-way inner join that combines data from the MANUFACTURERS, PRODUCTS, INVOICE, and CUSTOMERS tables.

SQL Code

```
PROC SQL;
  SELECT products.prodname,
         products.prodtype,
         customers.custname,
         manufacturers.manuname
    FROM MANUFACTURERS,
         PRODUCTS,
         INVOICE,
         CUSTOMERS
      WHERE manufacturers.manunum = products.manunum  AND
            manufacturers.manunum = invoice.manunum   AND
            products.prodnum      = invoice.prodnum   AND
            invoice.custnum       = customers.custnum;
QUIT;
```

Results

```
                                  The SAS System

Product Name           Product Type  Customer Name                Manufacturer Name
-------------------------------------------------------------------------------------
Analog Cell Phone      Phone         La Mesa Computer Land        Global Comm Corp
Spreadsheet Software   Software      Vista Tech Center            Incredible Software
Dream Machine          Workstation   Alpine Technical Center      Cupid Computer
Database Software      Software      Jamul Hardware & Software    KPL Enterprises
```

7.6 Outer Joins

As the previous examples in this chapter have shown, an inner join disregards any rows where the search condition is not met. This differs significantly from the way an *outer join* groups tables. In contrast with an inner join, an outer join keeps rows that match the ON (search) condition, as well as preserving some or all of the unmatched data from one or both of the tables. Essentially, an outer join retains rows from one table even when they do not match rows in the second table. This distinction is critical because this is what truly differentiates an outer join from an inner join.

Next, an outer join is capable of processing a maximum of two tables at a time, whereas (under the SAS implementation) an inner join is able to process a maximum of 32 tables.

Another difference has to do with how you specify outer join syntax. The comma used to designate or delimit one table from the other in the FROM clause of inner joins is replaced with one of the following keywords: LEFT JOIN, RIGHT JOIN, or FULL JOIN in outer joins Additionally, the WHERE clause expression used to restrict what rows are kept in the result table is replaced with the *ON* keyword.

Finally, an outer join is considered to be an asymmetric join (Lorie, Raymond A. and Jean-Jacques Daudenarde, *SQL & Its Applications*, page 87). Unlike inner joins, an outer join does not select rows proportionally from its parts or tables.

7.6.1 Left Outer Joins

Let's look at how a left join is applied in a real-world situation. Suppose you want to see a list of all manufacturers, their city locations, manufacturer numbers, their product types, and product costs (if available) without leaving out those manufacturers that do not have products yet. This means that the MANUFACTURERS table (left table) acts as the master table having its rows preserved while the PRODUCTS table (right table) acts as the contributing table (subordinate table). The following left outer join example effectively retains those matched rows from both tables as well as those rows from the left table that have no match in the right table.

SQL Code

```
PROC SQL;
  SELECT manuname, manucity, manufacturers.manunum,
         products.prodtype, products.prodcost
    FROM MANUFACTURERS LEFT JOIN PRODUCTS      ❶
       ON manufacturers.manunum =      ❷
          products.manunum;
QUIT;
```

❶ The LEFT JOIN specification preserves all the rows in the left table (MANUFACTURERS) even when there are no matching rows in the right table (PRODUCTS).

❷ The ON clause acts as a WHERE clause to select the desired rows in the join results.

As the results from the left outer join illustrate, the rows in the left (MANUFACTURERS) table that match rows in the right (PRODUCTS) table are included in the result table. As a result, eight rows match as evidenced by the value assigned to product type and product cost. Additionally, two rows from the left table that do not match rows in the right table (based on the search condition) are also retained (bolded). Therefore, each row from the MANUFACTURERS table that does not have a matching value in the PRODUCTS table is added to the resulting virtual table, accompanied by null values in the product type and product cost columns.

Results

The SAS System

Manufacturer Name	Manufacturer City	Manufacturer Number	Product Type	Product Cost
Cupid Computer	Houston	111	Workstation	$3,200.00
Storage Devices Inc	San Mate	120	Workstation	$3,300.00
Global Comm Corp	San Diego	210	Phone	$175.00
Global Comm Corp	San Diego	210	Phone	$35.00
KPL Enterprises	San Diego	500	Software	$299.00
KPL Enterprises	San Diego	500	Software	$299.00
KPL Enterprises	San Diego	500	Software	$299.00
KPL Enterprises	San Diego	500	Software	$399.00
World Internet Corp	**Miami**	**600**		**.**
San Diego PC Planet	**San Diego**	**700**		**.**

7.6.1.1 Specifying a WHERE Clause

To provide greater subsetting capabilities as well as added flexibility, the SQL procedure also permits the specification of an optional WHERE clause in addition to an ON clause when constructing outer joins. The ability to specify a WHERE clause in conjunction with an ON clause permits greater control over the subsetting of rows. An example will help illustrate how a WHERE clause is used in an outer join. Suppose you want to limit the results from the previous left outer join to only those products costing less than $300. In this example, the left outer join syntax uses a WHERE clause to subset row results to nonmissing products that cost less than $300.

SQL Code

```
PROC SQL;
  SELECT manuname, manucity, manufacturers.manunum,
         products.prodtype, products.prodcost
    FROM MANUFACTURERS LEFT JOIN PRODUCTS
      ON manufacturers.manunum =
         products.manunum
        WHERE prodcost < 300 AND      ❶
              prodcost NE .;
QUIT;
```

❶ The optional WHERE clause specified in addition to an ON clause in an outer join further subsets the joined results.

Results

The SAS System

Manufacturer Name	Manufacturer City	Manufacturer Number	Product Type	❶ Product Cost
Global Comm Corp	San Diego	210	Phone	$175.00
Global Comm Corp	San Diego	210	Phone	$35.00
KPL Enterprises	San Diego	500	Software	$299.00
KPL Enterprises	San Diego	500	Software	$299.00
KPL Enterprises	San Diego	500	Software	$299.00

7.6.1.2 Specifying Aggregate Functions

Suppose you need to produce a monthly report consisting of a total invoice amount by manufacturer. An aggregate function can be specified with outer join syntax to perform a group computation using a GROUP BY clause. In the next example, a left join computes the total invoice amount for each manufacturer with a SUM function and GROUP BY clause.

SQL Code

```
PROC SQL;
  SELECT manuname,
         SUM(invoice.invprice) AS Total_Invoice_Amt     ❶
             FORMAT=DOLLAR10.2
    FROM MANUFACTURERS LEFT JOIN INVOICE
      ON manufacturers.manunum =
         invoice.manunum
        GROUP BY MANUNAME;     ❷
QUIT;
```

❶ The SUM function computes the total invoice amount for each manufacturer.

❷ The GROUP BY clause groups all rows associated with a manufacturer into a single row.

The results show that manufacturers with no activity have a null or missing value in the aggregated Total_Invoice_Amt column.

Results

```
                    The SAS System
                                            ❶
        ❷                                  Total_
   Manufacturer Name                  Invoice_Amt
   ----------------------------------------------
   Cupid Computer                       $9,600.00
   Global Comm Corp                       $245.00
   KPL Enterprises                     $25,789.00
   San Diego PC Planet                          .
   Storage Devices Inc                          .
   World Internet Corp                  $1,598.00
```

7.6.2 Right Outer Joins

Right joins are similar to left joins, except the rows in the right (second) table are preserved. Consequently, the results will contain the rows of the symmetric join plus a row for each unmatched row in the right table. Nulls are automatically substituted for values from the left table. Suppose you want to see all manufacturers and their respective products. In the next example, a simple report containing products, product type, manufacturer number, and manufacturer name is produced from the PRODUCTS and MANUFACTURERS tables using a right outer join construct.

SQL Code

```
PROC SQL;
  SELECT prodname, prodtype,
         products.manunum, manuname
    FROM PRODUCTS RIGHT JOIN MANUFACTURERS     ❶
       ON products.manunum =
          manufacturers.manunum;
QUIT;
```

❶ The RIGHT JOIN specification preserves all the rows in the right table (MANUFACTURERS) even when there are no matching rows in the left table (PRODUCTS).

The results show that manufacturers appearing in the MANUFACTURERS table with no products listed in the PRODUCTS table have null or missing values in the Product Name, Product Type, and Manufacturer Number columns.

Note: To remove rows with missing values in the results, a WHERE clause could be specified.

Results

```
                              The SAS System
                                                                ❶
                                          Manufacturer    Manufacturer
Product Name              Product Type          Number    Name
------------------------------------------------------------------------
Dream Machine             Workstation              111    Cupid Computer
Business Machine          Workstation              120    Storage Devices Inc
Digital Cell Phone        Phone                    210    Global Comm Corp
Analog Cell Phone         Phone                    210    Global Comm Corp
Spreadsheet Software      Software                 500    KPL Enterprises
Graphics Software         Software                 500    KPL Enterprises
Wordprocessor Software    Software                 500    KPL Enterprises
Database Software         Software                 500    KPL Enterprises
                                                     .    World Internet Corp
                                                     .    San Diego PC Planet
```

7.6.3 Full Outer Joins

Full outer joins combine the power of left and right joins by preserving rows from both the left and right tables. Although a full join is not used as frequently as left join or right join constructs, it can be useful when information from both tables is missing. In the next example, a full outer join is specified to produce a report containing manufacturers with no products and products with no known manufacturers.

SQL Code

```
PROC SQL;
  SELECT prodname, prodtype,
         products.manunum, manuname
    FROM PRODUCTS FULL JOIN MANUFACTURERS     ❶
      ON products.manunum =
         manufacturers.manunum;
QUIT;
```

❶ The full join specification preserves all the rows in the left table (PRODUCTS) as well as all rows in the right table (MANUFACTURERS) even when there are no matching rows.

Results

The SAS System

Product Name	Product Type	Manufacturer Number ❶	Manufacturer Name
Dream Machine	Workstation	111	Cupid Computer
Business Machine	Workstation	120	Storage Devices Inc
Travel Laptop	Laptop	170	
Digital Cell Phone	Phone	210	Global Comm Corp
Analog Cell Phone	Phone	210	Global Comm Corp
Office Phone	Phone	220	
Spreadsheet Software	Software	500	KPL Enterprises
Graphics Software	Software	500	KPL Enterprises
Wordprocessor Software	Software	500	KPL Enterprises
Database Software	Software	500	KPL Enterprises
		.	World Internet Corp
		.	San Diego PC Planet

7.7 Subqueries

Now that we have seen how two or more tables can be combined in a join query, we turn our attention to another type of complex query known as a subquery. A subquery is a query expression that is nested within another query expression. Its purpose is to have the inner query produce a single value or multiple values that can then be passed into the outer query for processing. You achieve this by embedding a SELECT statement inside a WHERE clause of an outer query's SELECT statement, INSERT statement, DELETE statement, or HAVING clause.

Note: You should avoid nesting more than two subqueries deep because of the conceptual and processing complexities this introduces.

The typical subquery consists of a (inner) query combined inside the predicate of another (outer or main) query. When processed, the inner query passes a Boolean value to the outer query consisting of either *True* if it returns a minimum of one row or *False* if no rows are returned by the subquery. The results of the inner query are stored in a temporary results table and used as input to the main query. Our exploration of subqueries will involve using them with comparison operators, the IN predicate, and the ANY and ALL keywords, and will conclude with a look at a special type of subquery called a correlated subquery.

7.7.1 Alternate Approaches to Subqueries

A subquery is a very useful construct, especially when information from multiple tables needs to be interrelated. Unfortunately, a subquery is not always easy to construct and may even be more difficult to understand. So before constructing every table relation with a subquery, consider your options carefully.

When all the information is available in a single table, a simple query is probably all that needs to be constructed. Suppose you want to produce a report consisting of the invoice information for Global Comm Corp. Let's further assume you know the specific manufacturer number for Global Comm Corp as well. Knowing this means that you don't have to go into the MANUFACTURERS table to find it. In the next example, a simple query is constructed to retrieve all invoice information from the INVOICE table.

Simple Query

```
PROC SQL;
  SELECT *
    FROM INVOICE
      WHERE manunum = 210;
QUIT;
```

Results

The SAS System

Invoice Number	Manufacturer Number	Customer Number	Invoice Quantity - Units Sold	Invoice Unit Price	Product Number
1003	210	101	7	$245.00	2101

But what if all the information is not in a single table? And what if the manufacturer number for Global Comm Corp is not known? As shown earlier, a join can be constructed just as easily as a subquery. Some users prefer joins to subqueries because they can be easier to understand as well as maintain. In fact, a join frequently performs better than a subquery. In the next example, the manufacturer number for Global Comm Corp is not known. Consequently, a simple inner join is needed to retrieve all related rows from the MANUFACTURERS and INVOICE tables for Global Comm Corp.

Simple Join

```
PROC SQL;
  SELECT M.manunum, M.manuname, I.invnum,
         I.invqty, I.invprice
    FROM MANUFACTURERS M, INVOICE I
      WHERE M.manunum = I.manunum AND
            M.manuname = 'Global Comm Corp';
QUIT;
```

Results

The SAS System

Manufacturer Number	Manufacturer Name	Invoice Number	Invoice Quantity - Units Sold	Invoice Unit Price
210	Global Comm Corp	1003	7	$245.00

7.7.2 Passing a Single Value with a Subquery

Now let's see how a subquery could be constructed to provide the same results as with the join. As before, suppose you want to pull all the invoices for the manufacturer Global Comm Corp but only know the manufacturer name (or at least part of the name), but not the manufacturer number (MANUNUM). The following subquery uses an = (equal sign) in its outer query WHERE clause to accomplish this.

Since the manufacturer number is not known, a subquery is constructed to first search for it in the MANUFACTURERS table. Actually, the subquery approach is more versatile than the previous query approach, because it does not require a unique manufacturer number, which is often more difficult to remember than names. It also enables quick searches even if the manufacturer number changes for a given manufacturer.

When the entire query is executed, SQL first evaluates the inner query (or subquery) within the outer query's WHERE clause. It executes the inner query the same way as if it were a stand-alone query. It searches the MANUFACTURERS table for any row where the manufacturer name equals the character string Global Comm Corp and then pulls the MANUNUM values for this row. SQL then substitutes the derived MANUNUM value of 210 from the inner query inside the predicate of the main query (outer query). As a result of this substitution, the SQL query looks identical to the query mentioned previously.

SQL Code

```
PROC SQL;
  SELECT invnum, INVOICE.manunum, custnum, invqty, invprice,
      prodnum
    FROM INVOICE
      WHERE manunum =
        (SELECT manunum      ❶
          FROM MANUFACTURERS
            WHERE manuname = 'Global Comm Corp');
      WHERE INVOICE.manunum=MANUFACTURERS.manunum;
QUIT;
```

Result of Inner Query

```
PROC SQL;
  SELECT *
    FROM INVOICE
      WHERE manunum = 210;       ❷
QUIT;
```

❶ PROC SQL evaluates the inner query within the outer query's WHERE clause to search for the manufacturer number for manufacturer Global Comm Corp.

❷ The resulting query after substituting the derived manufacturer number value from the inner query evaluates to a single value and is then executed as the main (outer) query.

Results

The SAS System

Invoice Number	❷ Manufacturer Number	Customer Number	Invoice Quantity - Units Sold	Invoice Total Price	Product Number
1003	210	101	7	$245.00	2101

Let's look at another subquery. Suppose you want to retrieve the invoice from the INVOICE table for the manufacturer that manufactures the Dream Machine workstation. The following subquery (inner query) extracts the product number (PRODNUM) associated with the Dream Machine and passes the single value to the outer query for processing.

SQL Code

```
PROC SQL;
  SELECT invnum, manunum, custnum, invqty, invprice,
       INVOICE.prodnum
    FROM INVOICE
        (SELECT prodnum      ❶
          FROM PRODUCTS
            WHERE prodname LIKE 'Dream%')
      WHERE INVOICE.prodnum=PRODUCTS.prodnum;
QUIT;
```

Result of Inner Query

```
PROC SQL;
   SELECT *
     FROM INVOICE
       WHERE prodnum = 1110;      ❷
 QUIT;
```

❶ PROC SQL evaluates the inner query within the outer query's WHERE clause to search for the product number for the product Dream Machine.

❷ The resulting inner query after substituting the derived product number value evaluates to a single value and is then executed as the main (outer) query.

Results

The SAS System

Invoice Number	Manufacturer Number	Customer Number	Invoice Quantity - Units Sold	Invoice Unit Price	❷ Product Number
1004	111	501	3	$9,600.00	1110

It is fortunate that our subquery in the previous example passed only one row or value to the main (outer) query. Had it returned more than one value from the PRODUCTS table, it would have made it impossible for the SQL to evaluate the condition as true or false and would have produced an error in the outer query. Let's look at another example where more than one value is returned by the subquery.

In the next example, more than one row is returned by the inner query making it impossible for the main query to evaluate as true or false. As a result, an error is produced and the subquery does not execute. In general, it is best to avoid using the = (equal sign) and other comparison operators (<, >, <=, >=, and <>) in a subquery expression, unless you know in advance that the result of the subquery is a table with a single row of data (although it may not always be possible to know this beforehand). In a later section ("Passing More Than One Row with a Subquery"), you will see this problem alleviated by using the IN predicate.

SQL Code

```
PROC SQL;
  SELECT *
    FROM INVOICE
      WHERE manunum =
        (SELECT manunum
          FROM MANUFACTURERS
            WHERE UPCASE(manucity) LIKE 'SAN DIEGO%');
QUIT;
```

SAS Log Result

```
     PROC SQL;
       SELECT *
         FROM INVOICE
           WHERE manunum =
              (SELECT manunum
                FROM MANUFACTURERS
                  WHERE UPCASE(manucity) LIKE 'SAN DIEGO%');
ERROR: Subquery evaluated to more than one row.
     QUIT;
NOTE: The SAS System stopped processing this step because of errors.
NOTE: PROCEDURE SQL used:
      real time           0.00 seconds
```

Let's look at another subquery example that uses the comparison operator < (less than). A summary function specified in an inner query forces a single row to result. In the next example, the subquery uses the AVG summary (aggregate) function to determine which products (based on their invoice quantities) were purchased in lower quantities than the average product purchase.

SQL Code

```
PROC SQL;
  SELECT prodnum, invnum, invqty, invprice
    FROM INVOICE
      WHERE invqty <
        (SELECT AVG(invqty)       ❶
          FROM INVOICE);
QUIT;
```

Result of Inner Query

```
PROC SQL;
  SELECT prodnum, invnum, invqty, invprice
    FROM INVOICE
      WHERE invqty < 4.285714;     ❷
QUIT;
```

❶ PROC SQL evaluates the inner query within the outer query's WHERE clause to produce an average invoice quantity.

❷ The resulting inner query passes the derived average invoice quantity of 4.285714 as a single value to the main (outer) query for execution.

Results

The SAS System

Product Number	Invoice Number	❷ Invoice Quantity - Units Sold	Invoice Unit Price
6001	1002	2	$1,598.00
1110	1004	3	$9,600.00
5002	1005	2	$798.00
6000	1006	4	$396.00

7.7.3 Passing More Than One Row with a Subquery

To prevent the problem associated with passing more than one value to the main (outer) query, you can specify the IN predicate in a subquery. Similar to the IN operator in the DATA step, the IN predicate permits the SQL procedure to pass multiple row values from the (inner) subquery to the main (outer) query without producing an error.

Note: PROC SQL does not permit a subquery to select more than one column. The next example shows how multiple row values are passed from the subquery to the main (outer) query using the IN predicate for San Diego manufacturers.

SQL Code

```
PROC SQL;
  SELECT *
    FROM INVOICE
      WHERE manunum IN        ❶
        (SELECT manunum
          FROM MANUFACTURERS
            WHERE UPCASE(manucity) LIKE 'SAN DIEGO%');      ❷
QUIT;
```

Result of Inner Query

```
PROC SQL;
  SELECT prodnum, invnum, invqty, invprice
    FROM INVOICE
      WHERE manunum IN (210, 500, 700);     ❸
QUIT;
```

❶ PROC SQL's IN predicate is specified in the outer query to process a list of values that are passed from the inner query.

❷ PROC SQL evaluates the inner query within the outer query's WHERE clause to produce a list of manufacturer numbers for San Diego manufacturers.

❸ The resulting inner query passes multiple row values to the main (outer) query for execution.

Result

The SAS System

Invoice Number	❸ Manufacturer Number	Customer Number	Invoice Quantity - Units Sold	Invoice Total Price	Product Number
1001	500	201	5	$1,495.00	5001
1003	210	101	7	$245.00	2101
1005	500	801	2	$798.00	5002
1006	500	901	4	$396.00	6000
1007	500	401	7	$23,100.00	1200

7.7.4 Comparing a Set of Values

A subquery can have multiple values returned for a single column to the outer query. But there are special keywords that permit comparison operators to be used in subqueries to process multiple values. The special keywords ANY and ALL can be used to compare a set of values returned by a subquery. Let's see how these keywords work.

Suppose you want to view the products whose inventory quantity is greater than or equal to the lowest average inventory quantity. The following example illustrates a subquery with the *ANY* keyword specified in the WHERE clause of the main query expression. When *ANY* is specified, the entire WHERE clause is true if the subquery returns at least one value.

SQL Code

```
PROC SQL;
  SELECT manunum, prodnum, invqty, invprice
    FROM INVOICE
      WHERE invprice GE ANY      ❶
        (SELECT invprice
          FROM INVOICE
            WHERE prodnum IN (5001,5002));     ❷
QUIT;
```

Result of Inner Query

```
PROC SQL;
  SELECT manunum, prodnum, invqty, invprice
    FROM INVOICE
      WHERE invprice > ANY ($1,495.,$798.);    ❸
QUIT;
```

❶ PROC SQL retrieves any invoices from the outer query where the invoice price is greater than or equal to the row values passed from the inner query.

❷ The WHERE clause of the inner query retrieves any invoice prices for product numbers 5001 and 5002 and passes them to the outer query.

❸ The resulting inner query passes multiple row values to the main (outer) query for execution.

Results

The SAS System

Manufacturer Number	Product Number	Invoice Quantity - Units Sold	Invoice Unit Price
500	5001	5	$1,495.00
600	6001	2	$1,598.00
111	1110	3	$9,600.00
500	5002	2	$798.00
500	1200	7	$23,100.00

The ALL keyword works very differently from the ANY keyword. When you specify ALL before a subquery expression, the subquery is true only if the comparison is true for values returned by the subquery. For example, suppose you want to view the products whose inventory quantity is less than the average inventory quantity.

SQL Code

```
PROC SQL;
  SELECT manunum, prodnum, invqty, invprice
    FROM INVOICE
      WHERE invprice < ALL       ❶
        (SELECT invprice
          FROM INVOICE
            WHERE prodnum IN (5001,5002));     ❷
QUIT;
```

Result of Inner Query

```
PROC SQL;
  SELECT manunum, prodnum, invqty, invprice
    FROM INVOICE
      WHERE invprice < ALL ($1,495.,$798.);    ❸
QUIT;
```

❶ PROC SQL retrieves all invoices from the outer query where the invoice price is less than the row values passed from the inner query.

❷ The WHERE clause of the inner query retrieves all invoice prices for product numbers 5001 and 5002 and passes them to the outer query.

❸ The resulting inner query passes multiple row values to the main (outer) query for execution.

Results

```
                         The SAS System

                                     Invoice
                                    Quantity
      Manufacturer   Product    - Units     Invoice Unit
            Number    Number       Sold            Price
      ----------------------------------------------------
               210      2101          7          $245.00
               500      6000          4          $396.00
```

7.7.5 Correlated Subqueries

In the subquery examples shown so far, the subquery (inner query) operates independently from the main (outer) query. Essentially the subquery's results are evaluated and used as input to the main (outer) query. Although this is a common way subqueries execute, it is not the only way. SQL also permits a subquery to accept one or more values from its outer query. Once the subquery executes, the results are then passed to the outer query. Subqueries of this variety are called *correlated subqueries*. The ability to construct subqueries in this manner provides a powerful extension to SQL.

The difference between the subqueries discussed earlier and correlated subqueries is in the way the WHERE clause is constructed. Correlated subqueries relate a column in the subquery with a column in the outer query to determine the rows that match or in certain cases don't match the expression. Suppose, for example, that we want to view products in the PRODUCTS table that do not appear in the INVOICE table. One way to do this is to construct a correlated subquery.

In the next example, the subquery compares the product number column in the PRODUCTS table with the product number column in the INVOICE table. If at least one match is found (the product appears in both the PRODUCTS and INVOICE tables) then the resulting table from the subquery will not be empty, and the NOT EXISTS condition will be false. However, if no matches are found, then the subquery returns an empty table resulting in the NOT EXISTS condition being true, causing the product number, product name, and product type of the current row in the main (outer) query to be selected.

SQL Code

```
PROC SQL;
  SELECT prodnum, prodname, prodtype
    FROM PRODUCTS
      WHERE NOT EXISTS       ❶
        (SELECT *
          FROM INVOICE
            WHERE PRODUCTS.prodnum = INVOICE.prodnum);     ❷
QUIT;
```

❶ The (inner) subquery receives its value(s) from the main (outer) query. With the value(s), the subquery runs and passes the results back to the main query where the WHERE clause and the NOT EXISTS condition are processed.

❷ The inner query selects matching product and invoice information and passes it to the outer query.

Results

```
                    The SAS System

 Product
 Number    Product Name                   Product Type
 ______________________________________________________
   1700    Travel Laptop                  Laptop
   2102    Digital Cell Phone             Phone
   2200    Office Phone                   Phone
   5003    Wordprocessor Software         Software
   5004    Graphics Software              Software
```

Correlated subqueries are useful for placing restrictions on the results of an entire query with a HAVING clause (or, when combined with a GROUP BY clause, of an entire group). Suppose you want to know which manufacturers have more than one invoiced product.

In the next example, the subquery compares the manufacturer number in the PRODUCTS table with the manufacturer number in the INVOICE table. A HAVING clause and a COUNT function are specified to select all manufacturers with two or more invoices. Because an aggregate (summary) function is used in an optional HAVING clause, a GROUP BY clause is not needed to select the manufacturers with two or more invoices. An EXISTS condition is specified in the outer query's WHERE clause to capture only those manufacturers matching the subquery.

SQL Code

```
PROC SQL;
  SELECT prodnum, prodname, prodtype
    FROM PRODUCTS
      WHERE EXISTS      ❶
        (SELECT *
          FROM INVOICE
            WHERE PRODUCTS.manunum = INVOICE.manunum
              HAVING COUNT(*) > 1);     ❷
QUIT;
```

❶ The (inner) subquery receives its value(s) from the main (outer) query. With the value(s), the subquery runs and passes the results back to the main query where the WHERE clause and the EXISTS condition are processed.

❷ The inner query specifies a HAVING clause in order to subset manufacturers with two or more invoices.

Results

```
                          The SAS System

      Product
       Number    Product Name                   Product Type
     ----------------------------------------------------------
         5001    Spreadsheet Software           Software
         5002    Database Software              Software
         5003    Wordprocessor Software         Software
         5004    Graphics Software              Software
```

7.8 Set Operations

Now that we have seen how tables are combined with join queries and subqueries, we turn our attention to another type of complex query. The SQL procedure provides users with several table operators: INTERSECT, UNION, OUTER UNION, and EXCEPT, commonly referred to as set operators. In contrast to joins and subqueries where query results are combined horizontally, the purpose of each set operator is to combine or concatenate query results vertically.

7.8.1 Accessing Rows from the Intersection of Two Queries

The INTERSECT operator creates query results consisting of all the unique rows from the intersection of the two queries. Put another way, the intersection of two queries (A and B) is represented by C, indicating that the rows that are produced occur in both A and in B. As the following figure shows, the intersection of both queries is represented in the shaded area (C).

Figure 7.1. Intersection of Two Queries

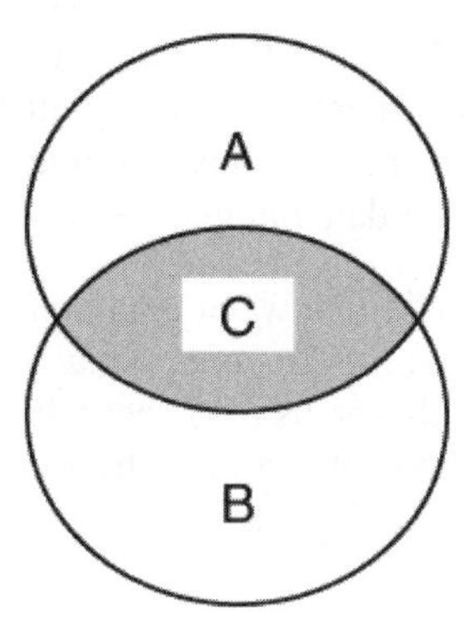

To see all products that cost less than $300.00 and product types classified as "phone", you could construct a simple query with a WHERE clause or specify the intersection of two separate queries. The next example illustrates a simple query that specifies a WHERE clause to display phones that cost less than $300.

SQL Code

```
PROC SQL;
  SELECT *
    FROM PRODUCTS
      WHERE prodcost < 300.00 AND
            prodtype = 'Phone';
  QUIT;
```

Results

The SAS System

Product Number	Product Name	Manufacturer Number	Product Type	Product Cost
2101	Analog Cell Phone	210	Phone	$35.00
2102	Digital Cell Phone	210	Phone	$175.00
2200	Office Phone	220	Phone	$130.00

The INTERSECT approach can be constructed to produce the same results as in the previous example. The INTERSECT process assumes that the tables in each query are structurally identical to each other. It overlays the columns from both queries based on position in the SELECT statement. Should you attempt to intersect two queries with different table structures, the process may fail due to differing column types, or the produced results may contain data integrity issues.

The most significant distinction between the two approaches, and one that may affect large table processing, is that the first query example (using the AND operator) takes less time to process: 0.05 seconds versus 0.17 seconds for the second approach (using the INTERSECT operator). The next example shows how the INTERSECT operator achieves the same result less efficiently.

SQL Code

```
PROC SQL;
  SELECT *        ❶
    FROM PRODUCTS
      WHERE prodcost < 300.00

  INTERSECT              ❷

  SELECT *        ❶
    FROM PRODUCTS
      WHERE prodtype = "Phone";
QUIT;
```

❶ It is assumed that the tables in both queries are structurally identical because the wildcard character "*" is specified in the SELECT statement.

❷ The INTERSECT operator produces rows common to both queries.

Results

The SAS System

Product Number	Product Name	Manufacturer Number	Product Type	Product Cost
2101	Analog Cell Phone	210	Phone	$35.00
2102	Digital Cell Phone	210	Phone	$175.00
2200	Office Phone	220	Phone	$130.00

7.8.2 Accessing Rows from the Combination of Two Queries

The UNION operator preserves all the unique rows from the combination of queries. The result is the same as if an OR operator is used to combine the results of each query. Put another way, the union of two queries (A and B) represents rows in A or in B or in both A and B. As illustrated in the figure below, the union represents the entire shaded area (A, B, and C).

Figure 7.2. Union of Two Queries

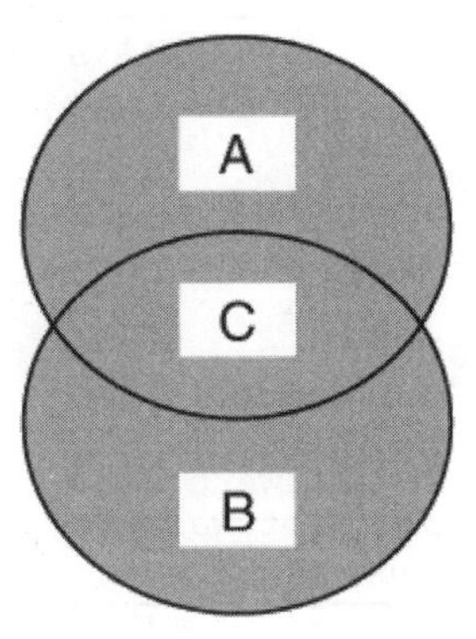

UNION automatically eliminates duplicate rows from the results, unless the ALL keyword is specified as part of the UNION operator. The column names assigned to the results are derived from the names in the first query.

In order for the union of two or more queries to be successful, each query must specify the same number of columns of the same or compatible types. Type compatibility means that column attributes are defined the same way. Because column names and attributes are derived from the first table, data types must be of the same type. The data types of the result columns are derived from the source table(s).

To see all products that cost less than $300.00 or products classified as a workstation, you have a choice between using OR as in the following query or UNION as in the next. As illustrated in the output from both queries, the results are identical no matter which query is used.

SQL Code

```
PROC SQL;
  SELECT *
    FROM PRODUCTS
      WHERE prodcost < 300.00 OR
            prodtype = "Workstation";
  QUIT;
```

Results

The SAS System

Product Number	Product Name	Manufacturer Number	Product Type	Product Cost
1110	Dream Machine	111	Workstation	$3,200.00
1200	Business Machine	120	Workstation	$3,300.00
2101	Analog Cell Phone	210	Phone	$35.00
2102	Digital Cell Phone	210	Phone	$175.00
2200	Office Phone	220	Phone	$130.00
5001	Spreadsheet Software	500	Software	$299.00
5003	Wordprocessor Software	500	Software	$299.00
5004	Graphics Software	500	Software	$299.00

In the next example, the UNION operator is specified to combine the results of both queries.

SQL Code

```
PROC SQL;
  SELECT *
    FROM PRODUCTS
      WHERE prodcost < 300.00

  UNION
```

```
  SELECT *
    FROM PRODUCTS
      WHERE prodtype = 'Workstation';
QUIT;
```

❶ The UNION operator combines the results of two queries.

Results

The SAS System

Product Number	Product Name	Manufacturer Number	Product Type	Product Cost
1110	Dream Machine	111	Workstation	$3,200.00
1200	Business Machine	120	Workstation	$3,300.00
2101	Analog Cell Phone	210	Phone	$35.00
2102	Digital Cell Phone	210	Phone	$175.00
2200	Office Phone	220	Phone	$130.00
5001	Spreadsheet Software	500	Software	$299.00
5003	Wordprocessor Software	500	Software	$299.00
5004	Graphics Software	500	Software	$299.00

7.8.3 Concatenating Rows from Two Queries

The OUTER UNION operator concatenates the results of two queries. As with a DATA step or PROC APPEND concatenation the results consist of rows combined vertically. Put another way, the outer union of two queries (A and B) represents all rows in both A and B with no overlap. As illustrated below, the outer union represents the entire shaded area (A and B).

Figure 7.3. Outer Union of Two Queries

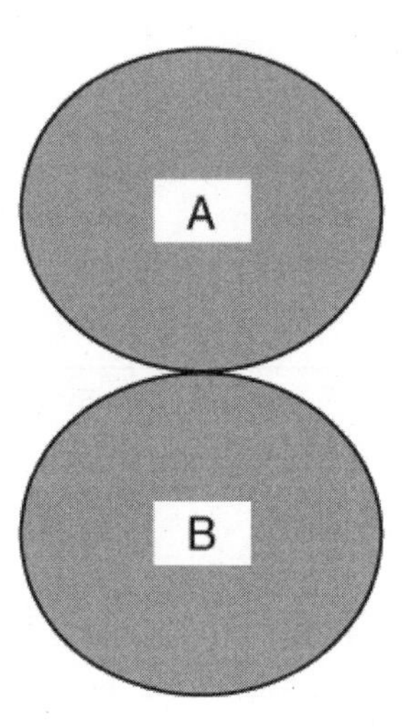

The next example concatenates the results of two queries. As illustrated in the output, the results show the rows from both queries are concatenated.

SQL Code

```
PROC SQL;
  SELECT prodnum, prodname, prodtype, prodcost
    FROM PRODUCTS

  OUTER UNION            ❶

  SELECT prodnum, prodname, prodtype, prodcost
    FROM PRODUCTS;
QUIT;
```

❶ The OUTER UNION operator concatenates the results of both queries.

Results

The SAS System

Product Type	Product Cost	Product Type	Product Cost
Workstation	$3,200.00		.
Workstation	$3,300.00		.
Laptop	$3,400.00		.
Phone	$35.00		.
Phone	$175.00		.
Phone	$130.00		.
Software	$299.00		.
Software	$399.00		.
Software	$299.00		.
Software	$299.00		.
	.	Workstation	$3,200.00
	.	Workstation	$3,300.00
	.	Laptop	$3,400.00
	.	Phone	$35.00
	.	Phone	$175.00
	.	Phone	$130.00
	.	Software	$299.00
	.	Software	$399.00
	.	Software	$299.00
	.	Software	$299.00

The OUTER UNION operator automatically concatenates rows from two queries with no overlap, unless the CORRESPONDING (CORR) keyword is specified as part of the operator. The column names assigned to the results are derived from the names in the first query. In the next example, the CORR keyword enables columns with the same name and attributes to be overlaid.

SQL Code

```
PROC SQL;
  SELECT prodnum, prodname, prodtype, prodcost
    FROM PRODUCTS

  OUTER UNION CORR          ❶

  SELECT prodnum, prodname, prodtype, prodcost
    FROM PRODUCTS;
QUIT;
```

❶ The OUTER UNION operator with the CORR keyword concatenates and overlays the results of both queries.

Results

The SAS System

Product Number	Product Name	Product Type	Product Cost
1110	Dream Machine	Workstation	$3,200.00
1200	Business Machine	Workstation	$3,300.00
1700	Travel Laptop	Laptop	$3,400.00
2101	Analog Cell Phone	Phone	$35.00
2102	Digital Cell Phone	Phone	$175.00
2200	Office Phone	Phone	$130.00
5001	Spreadsheet Software	Software	$299.00
5002	Database Software	Software	$399.00
5003	Wordprocessor Software	Software	$299.00
5004	Graphics Software	Software	$299.00
1110	Dream Machine	Workstation	$3,200.00
1200	Business Machine	Workstation	$3,300.00
1700	Travel Laptop	Laptop	$3,400.00
2101	Analog Cell Phone	Phone	$35.00
2102	Digital Cell Phone	Phone	$175.00

(continued on next page)

```
2200    Office Phone                 Phone          $130.00
5001    Spreadsheet Software         Software       $299.00
5002    Database Software            Software       $399.00
5003    Wordprocessor Software       Software       $299.00
5004    Graphics Software            Software       $299.00
```

7.8.4 Comparing Rows from Two Queries

The EXCEPT operator compares rows from two queries to determine the changes made to the first table that are not present in the second table. The result below shows new and changed rows in the first table that are not in the second table, but not rows that have been deleted from the second table. As illustrated in figure 7-4, the results of specifying the EXCEPT operator represent the shaded area (A) in the diagram.

Figure 7.4. Compare Two Tables to Determine Additions and Changes

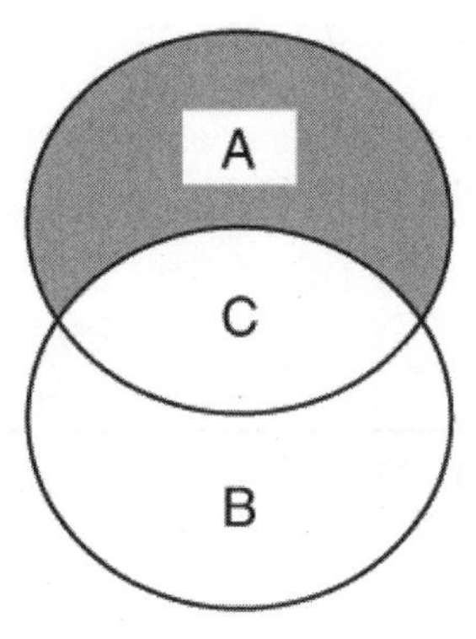

When working with two tables consisting of similar information, you can use the EXCEPT operator to determine new and modified rows. The EXCEPT operator compares rows in both tables to identify the rows existing in the first table but not in the second table. It also uniquely identifies rows that have changed from the first to the second tables. Columns are compared in the order they appear in the SELECT statement.

If the wildcard character "*" is specified in the SELECT statement, it is assumed that the tables are structurally identical to one another. Let's look at an example.

Suppose you have master and backup tables of the CUSTOMERS file, and you want to compare them to determine the new and changed rows. The EXCEPT operator as illustrated in the next example returns all new or changed rows from the CUSTOMERS table that do not appear in the CUSTOMERS_BACKUP table. As illustrated by the output, three new customer rows are added to the CUSTOMERS table that had not previously existed in the CUSTOMERS_BACKUP table.

SQL Code

```
PROC SQL;
  SELECT *
    FROM CUSTOMERS

  EXCEPT          ❶

  SELECT *
    FROM CUSTOMERS_BACKUP;
QUIT;
```

❶ The EXCEPT operator compares rows in both tables to identify the rows existing in the first table but not the second table.

Results

```
                         The SAS System

 Customer
   Number   Customer Name                 Customer's Home City
 ------------------------------------------------------------
     1302   Software Intelligence Cor     Spring Valley
     1901   Shipp Consulting              San Pedro
     1902   Gupta Programming             Simi Valley
```

7.9 Summary

1. When one or more relationships or connections between disparate pieces of data are needed, the PROC SQL join construct is used (see section 7.3.1).

2. You use a join to relate one table with another through a process known as column matching (see section 7.3.2).

3. You can assign table aliases to tables to minimize the number of keystrokes when referencing a table in a join query (see section 7.5.4).

4. When a query is placed inside the predicate of another query, it is called a subquery. Put another way, a subquery is a SELECT statement that is embedded in the WHERE clause of another SELECT statement (see section 7.7).

5. The IN predicate permits PROC SQL to pass multiple values from the subquery to the main query without producing an error (see section 7.7.3).

6. A subquery can also be constructed to evaluate multiple times, once for each row of data accessed by the main (outer) query (see section 7.7.4).

7. The INTERSECT operator creates an output table consisting of all the unique rows from the intersection of two query expressions (see section 7.8.1).

8. The UNION operator creates an output table consisting of all the unique rows from the combination of query expressions (see section 7.8.2).

Chapter 8

Working with Views

8.1 Introduction

In previous chapters, the examples assumed that each table had a physical existence, that is, the data stored in each table occupied storage space. In this chapter we turn our attention to a different type of table structure that has no real physical existence. This structure, known as a virtual table or view, offers users and programmers an incredible amount of flexibility and control. This makes views an ideal way to look at data from a variety of perspectives and according to different users' needs. Unlike tables, views store no data and have only a "virtual" existence. You will learn how to create, access, and delete views as you examine the many examples in this chapter.

8.2 Views — Windows to Your Data

Views are one of the more powerful features available in the SQL procedure. They are commonly referred to as "virtual tables" to distinguish them from base tables. The simple difference is that views are not tables, but files consisting of executable instructions. As a query, a view appears to behave as a table with one striking difference — it does not store any data. When referenced, a view produces results just like a table does. So how does a view get its data? Views access data from one or more underlying tables (base tables) or other views, provide you with your own personal access to data, and can be used in DATA steps as well as by SAS procedures.

Views offer improved control, manageability, and security in dynamic environments where data duplication or data redundancy, logic complexities, and data security are an issue. When used properly, views enable improved change control by providing enhanced data accessibility, hiding certain columns from unauthorized users, while enabling improved maintainability.

Data references are coded one time and, once a view is tested, can be conveniently stored in common and shareable libraries making them accessible for all to use. Views ensure that the most current input data is accessed and prevent the need for replicating partial or complete copies of the input data. As a means of shielding users from complex logic constructs, views can be designed to look as though a database were designed specifically for a single user as well as for a group of users, each with differing needs.

Views are also beneficial when queries or subqueries are repeated a number of times throughout an application. In these situations the addition of a view enables a change to be made only once, improving a your productivity through a reduction in time and resources. The creation of view libraries should be considered so users throughout an organization have an easily accessible array of productivity routines as they would a macro.

8.2.1 What Views Aren't

Views are not tables, but file constructs containing compiled code that access one or more underlying tables. Because views do not physically store data, they are referred to as "virtual" tables. Unlike tables, views do not physically contain or store rows of data. Views, however, do have a physical presence and take up space. Storage demands for views are minimal because the only portion saved is the SELECT statement or query itself. Tables, on the other hand, store one or more rows of data and their attributes within their structure.

Views are created with the CREATE VIEW statement while tables are created with the CREATE TABLE statement. Because you use one or more underlying tables to create a virtual (derived) table, views provide you with a powerful method for accessing data sources.

Although views have many unique and powerful features, they also have pitfalls. First, views generally take longer to process than tables. Each time a view is referenced, the current underlying table or tables are accessed and processed. Because a view is not physically materialized until it is accessed, higher utilization costs are typically involved, particularly for larger views. Next, views cannot create indexes on the underlying base tables. This can make it more difficult to optimize views.

8.2.2 Types of Views

Views can be designed to achieve a number of objectives:

- Referencing a single table
- Producing summary data across a row
- Concealing sensitive information
- Creating updatable views
- Grouping data based on summary functions or a HAVING clause
- Using set operators
- Combining two or more tables in join operations
- Nesting one view within another

As a way of distinguishing the various types of views, Joe Celko introduced a classification system based on the type of SELECT statement used (see *SQL for Smarties: Advanced SQL Programming*).

To help you understand the different view types, this chapter describes and illustrates view construction as well as how they can be used. A view can also have the characteristics of one or more view types, thereby being classified as a hybrid. A hybrid view, for example, could be designed to reference two or more tables, perform updates, and contain complex computations. The table below presents the different view types along with a brief description of their purpose.

A Description of the Various View Types

Types of Views	Description
Single-table view	a single-table view references a single underlying (base) table. It is the most common type of view. Selected columns and rows can be displayed or hidden depending on need.
Calculated column views	a calculated column view provides summary data across a row.
Read-only view	a read-only view prevents data from being updated (as opposed to updatable views) and is used to display data only. This also serves security purposes for the concealment of sensitive information.
Updatable view	an updatable view adds (inserts), modifies, or deletes rows of data.
Grouped view	a grouped view uses query expressions based on a query with a GROUP BY clause.
Set operation view	a set operation view includes the union of two tables, the removal of duplicate rows, the concatenation of results, and the comparison of query results.
Joined view	a joined view is based on the joining of two or more base tables. This type of view is often used in table-lookup operations to expand (or translate) coded data into text.
Nested view	a nested view is based on one view being dependent on another view such as with subqueries.
Hybrid view	an integration of one or more view types for the purpose of handling more complex tasks.

8.2.3 Creating Views

You use the CREATE VIEW statement in the SQL procedure to create a view. When the SQL processor sees the words CREATE VIEW, it expects to find a name assigned to the newly created view. The SELECT statement defines the names assigned to the view's columns as well as their order.

Views are often constructed so that the order of the columns is different from the base table. In the next example, a view is created with the columns appearing in a different order from the original MANUFACTURERS base table. The view's SELECT statement does not execute during this step because its only purpose is to define the view in the CREATE VIEW statement.

SQL Code

```
PROC SQL;
  CREATE VIEW MANUFACTURERS_VIEW AS
    SELECT manuname, manunum, manucity, manustat
      FROM MANUFACTURERS;
QUIT;
```

SAS Log Results

```
     PROC SQL;
       CREATE VIEW MANUFACTURERS_VIEW AS
         SELECT manuname, manunum, manucity, manustat
           FROM MANUFACTURERS;
NOTE: SQL view WORK.MANUFACTURERS_VIEW has been defined.
     QUIT;
NOTE: PROCEDURE SQL used:
      real time           0.44 seconds
```

When you create a view, you can create columns that are not present in the base table from which you built your view. That is, you can create columns that are the result of an operation (addition, subtraction, multiplication, etc.) on one or more columns in the base tables. You can also build a view using one or more unmodified columns of one or more base tables. Columns created this way are referred to as derived columns or calculated columns. In the next example, say we want to create a view consisting of the product name, inventory quantity, and inventory cost from the INVENTORY base table and a derived column of average product costs stored in inventory.

SQL Code

```
PROC SQL;
  CREATE VIEW INVENTORY_VIEW AS
    SELECT prodnum, invenqty, invencst,
           invencst/invenqty AS AverageAmount
      FROM INVENTORY;
QUIT;
```

SAS Log Results

```
     PROC SQL;
       CREATE VIEW INVENTORY_VIEW AS
         SELECT prodnum, invenqty, invencst,
                invencst/invenqty AS AverageAmount
           FROM INVENTORY;
NOTE: SQL view WORK.INVENTORY_VIEW has been defined.
     QUIT;
NOTE: PROCEDURE SQL used:
      real time           0.00 seconds
```

8.2.4 Displaying a View's Contents

You would expect the CONTENTS procedure to display information about the physical characteristics of a SAS data library and its tables. But what you may not know is that the CONTENTS procedure can also be used to display information about a view. The output generated from the CONTENTS procedure shows that the view contains no rows (observations) by displaying a missing value in the Observations field and a member type of View. The engine used is the SQLVIEW. The following example illustrates the use of the CONTENTS procedure in the Windows environment to display the INVENTORY_VIEW view's contents.

SQL Code

```
PROC CONTENTS DATA=INVENTORY_VIEW;
RUN;
```

SAS Output Results

```
                              The SAS System

                          The CONTENTS Procedure

Data Set Name   WORK.INVENTORY_VIEW                Observations            .
Member Type     VIEW                               Variables               4
Engine          SQLVIEW                            Indexes                 0
Created         Wednesday, August 18, 2004         Observation Length      32
Last Modified   Wednesday, August 18, 2004         Deleted Observations    0
Protection                                         Compressed              NO
Data Set Type                                      Sorted                  NO
Label
Data Representation  Default
Encoding             Default

      -----Alphabetic List of Variables and Attributes-----

#  Variable        Type   Len   Flags   Format       Label
______________________________________________________________________
4  AverageAmount   Num      8   P--
3  invencst        Num      6   -C-     DOLLAR10.2   Inventory Cost
2  invenqty        Num      3   -C-                  Inventory Quantity
1  prodnum         Num      3   ---                  Product Number
```

8.2.5 Describing View Definitions

Because views consist of partially compiled executable statements, ordinarily you will not be able to read the code in a view definition. However, the SQL procedure provides a statement to inspect the contents of the executable instructions (stored query expression) contained within a view definition. Without this capability, a view's underlying instructions (PROC SQL code) would forever remain a mystery and would make the ability to modify or customize the query expressions next to impossible. Whether your job is to maintain or customize a view, the DESCRIBE VIEW statement is the way you review the statements that make up a view. Let's look at how a view definition is described.

The next example shows the DESCRIBE VIEW statement being used to display the INVENTORY_VIEW view's instructions. It should be noted that results are displayed in the SAS log, not in the Output window.

SQL Code

```
PROC SQL;
   DESCRIBE VIEW INVENTORY_VIEW;
QUIT;
```

SAS Log Results

```
      PROC SQL;
        DESCRIBE VIEW INVENTORY_VIEW;
 NOTE: SQL view WORK.INVENTORY_VIEW is defined as:
         select prodnum, invenqty, invencst,
                invencst/invenqty as CostQty_Ratio
           from INVENTORY;
      QUIT;
 NOTE: PROCEDURE SQL used:
       real time           0.05 seconds
```

8.2.6 Creating and Using Views in the SAS System

Views are accessed the same way as tables. The SQL procedure permits views to be used in SELECT queries, subsets, joins, other views, and DATA and PROC steps. Views can reference other views (as will be seen in more detail in a later section), but the referenced views must ultimately reference one or more existing base tables.

The only thing that cannot be done is to create a view from a table or view that does not already exist. When this is attempted, an error message is written in the SAS log indicating that the view is being referenced recursively. An error occurs because the view being referenced directly (or indirectly) by it cannot be located or opened successfully. The next example shows the error that occurs when a view called NO_CAN_DO_VIEW is created from a non-existing view by the same name in a SELECT statement FROM clause.

SQL Code

```
PROC SQL;
  CREATE VIEW NO_CAN_DO_VIEW AS
    SELECT *
      FROM NO_CAN_DO_VIEW;

  SELECT *
    FROM NO_CAN_DO_VIEW;
QUIT;
```

SAS Log Results

```
   PROC SQL;
     CREATE VIEW NO_CAN_DO_VIEW AS
       SELECT *
         FROM NO_CAN_DO_VIEW;
NOTE: SQL view WORK.NO_CAN_DO_VIEW has been defined.
     SELECT *
       FROM NO_CAN_DO_VIEW;
ERROR: The SQL View WORK.NO_CAN_DO_VIEW is referenced recursively.
   QUIT;
NOTE: The SAS System stopped processing this step because of errors.
NOTE: PROCEDURE SQL used (Total process time):
      real time           0.00 seconds
      cpu time            0.01 seconds
```

8.2.7 Views and SAS Procedures

In most cases but not all, views can be used just as input SAS data sets to the universe of available SAS procedures. In the first example, the INVENTORY_VIEW view is used as input to the MEANS procedure to produce simple univariate descriptive statistics for numeric variables. Accessing the INVENTORY_VIEW view is different from accessing the INVENTORY table because the view's internal compiled executable statements are processed providing current data from the underlying table to the view itself. The view statements and the statements and options from the MEANS procedure determine what information is produced.

The next example uses the INVENTORY_VIEW view as input to the MEANS procedure to produce simple univariate descriptive statistics for numeric variables. Accessing the INVENTORY_VIEW view is different from accessing the INVENTORY table because the view derives and provides current data from the underlying table to the view itself. The view statements and the statements and options from the MEANS procedure determine what information is produced.

SAS Code

```
PROC MEANS DATA=INVENTORY_VIEW;
  TITLE1 'Inventory Statistical Report';
  TITLE2 'Demonstration of a View used in PROC MEANS';
RUN;
```

SAS Log Results

```
     PROC MEANS DATA=INVENTORY_VIEW;
       TITLE1 'Inventory Statistical Report';
       TITLE2 'Demonstration of a View used in PROC MEANS';
     RUN;
NOTE: There were 7 observations read from the dataset WORK.INVENTORY.
NOTE: There were 7 observations read from the dataset
      WORK.INVENTORY_VIEW.
NOTE: PROCEDURE MEANS used:
      real time           0.32 seconds
```

Results

```
                        Inventory Statistical Report
                  Demonstration of a View used in PROC MEANS

                            The MEANS Procedure

  Variable     Label                   N         Mean      Std Dev         Minimum
  --------------------------------------------------------------------------------
  prodnum      Product Number          7      3974.43      1763.50         1110.00
  invenqty     Inventory Quantity      7   10.0000000    7.5055535       2.0000000
  invencst     Inventory Cost          7     11357.14     17866.72     900.0000000
  AverageAmount                        7  917.1428571      1121.71      70.0000000
  --------------------------------------------------------------------------------

               Variable         Label                        Maximum
               -----------------------------------------------------
               prodnum          Product Number               5004.00
               invenqty         Inventory Quantity        20.0000000
               invencst         Inventory Cost              45000.00
               AverageAmount                                 2800.00
               -----------------------------------------------------
```

The next example uses the INVENTORY_VIEW view as input to the PRINT procedure to produce a detailed listing of the values contained in the underlying base table.

Note: It is worth noting that, as with all procedures, all procedure options and statements are available by views.

SAS Code

```
PROC PRINT DATA=INVENTORY_VIEW N NOOBS UNIFORM;
  TITLE1 'Inventory Detail Listing';
  TITLE2 'Demonstration of a View used in PROC PRINT';
  format AverageAmount dollar10.2;
RUN;
```

SAS Log Results

```
    PROC PRINT DATA=INVENTORY_VIEW N NOOBS UNIFORM;
      TITLE1 'Inventory Detail Listing';
      TITLE2 'Demonstration of a View used in a Procedure';
      format AverageAmount dollar10.2;
    RUN;
NOTE: There were 7 observations read from the dataset WORK.INVENTORY.
NOTE: There were 7 observations read from the dataset
      WORK.INVENTORY_VIEW.
NOTE: PROCEDURE PRINT used:
      real time           0.04 seconds
```

Results

```
                  Inventory Detail Listing
          Demonstration of a View used in PROC PRINT

                                                   Average
        prodnum     invenqty        invencst        Amount

           1110           20      $45,000.00     $2,250.00
           1700           10      $28,000.00     $2,800.00
           5001            5       $1,000.00       $200.00
           5002            3         $900.00       $300.00
           5003           10       $2,000.00       $200.00
           5004           20       $1,400.00        $70.00
           5001            2       $1,200.00       $600.00

                             N = 7
```

8.2.8 Views and DATA Steps

As we have already seen, views can be used as input to SAS procedures as if they were data sets. You will now see that views are a versatile component that can be used in a DATA step as well. This gives you a controlled way of using views to access tables of data in custom report programs. The next example uses the INVENTORY_VIEW view as input to the DATA step as if it were a SAS base table. Notice that the KEEP= data set option reads only two of the variables from the INVENTORY_VIEW view.

SAS Code

```
DATA _NULL_;
  SET INVENTORY_VIEW (KEEP=PRODNUM AVERAGEAMOUNT);
  FILE PRINT HEADER=H1;
  PUT @10 PRODNUM
      @30 AVERAGEAMOUNT DOLLAR10.2;
RETURN;
H1: PUT @9 'Using a View in a DATA Step'
     /// @5 'Product Number'
     @26 'Average Amount';
RETURN;
RUN;
```

SAS Log Results

```
     DATA _NULL_;
       SET INVENTORY_VIEW (KEEP=PRODNUM AVERAGEAMOUNT);
       FILE PRINT HEADER=H1;
       PUT @10 PRODNUM
           @30 AVERAGEAMOUNT DOLLAR10.2;
     RETURN;
     H1: PUT @9 'Using a View in a DATA Step'
         /// @5 'Product Number'
             @26 'Average Amount';
     RETURN;
     RUN;
NOTE: 11 lines were written to file PRINT.
NOTE: There were 7 observations read from the dataset WORK.INVENTORY.
NOTE: There were 7 observations read from the dataset
      WORK.INVENTORY_VIEW.
NOTE: DATA statement used:
     real time           0.00 seconds
```

Output

```
              Using a View in a DATA Step


       Product Number         Average Amount
            1110                   $2,250.00
            1700                   $2,800.00
            5001                     $200.00
            5002                     $300.00
            5003                     $200.00
            5004                      $70.00
            5001                     $600.00
```

8.3 Eliminating Redundancy

Data redundancy commonly occurs when two or more users want to see the same data in different ways. To prevent redundancy, organizations should create and maintain a master database environment rather than propagate one or more subsets of data among its users communities. The latter approach creates an environment that is not only problematic for the organization but for its users and customers.

Views provide a way to eliminate or, at least, reduce the degree of data redundancy. Rather than having the same data exist in multiple forms, views create a virtual and shareable database environment for all. Problems related to accessing and reporting outdated information, as well as table and program change control, are eliminated.

8.4 Restricting Data Access — Security

As data security issues grow increasingly important for organizations around the globe, views offer a powerful alternative in controlling or restricting access to sensitive information. Views, like tables, can prevent unauthorized users from accessing sensitive portions of data. This is important because security breeches pose great risks not only to an organization's data resources but to the customer as well.

Views can be constructed to show a view of data different from what physically exists in the underlying base tables. Specific columns can be shown while others are hidden. This helps prevent sensitive information such as salary, medical, or credit card data from getting into the wrong hands. Or a view can contain a WHERE clause with any degree of complexity to restrict what rows appear for a group of users while hiding other rows. In the next example, a view called SOFTWARE_PRODUCTS_VIEW is created that displays all columns from the original table except the product cost (PRODCOST) column and restricts all rows except "Software" from the PRODUCTS table.

SQL Code

```
PROC SQL;
  CREATE VIEW SOFTWARE_PRODUCTS_VIEW AS
    SELECT prodnum, prodname, manunum, prodtype
      FROM PRODUCTS
        WHERE UPCASE(PRODTYPE) IN ('SOFTWARE');
QUIT;
```

SAS Log Results

```
      PROC SQL;
      CREATE VIEW SOFTWARE_PRODUCTS_VIEW AS
        SELECT prodnum, prodname, manunum, prodtype
          FROM PRODUCTS
            WHERE UPCASE(PRODTYPE) IN ('SOFTWARE');
NOTE: SQL view WORK.SOFTWARE_PRODUCTS_VIEW has been defined.
   QUIT;
NOTE: PROCEDURE SQL used:
      real time           0.17 seconds
```

The SOFTWARE_PRODUCTS_VIEW view functions just as if it were a base table, although it contains no rows of data. All other columns with the exception of product cost (PRODCOST) are inherited from the selected columns in the PRODUCTS table. A view determines what columns and rows are processed from the underlying table and, optionally, the SELECT query referencing the view can provide additional criteria during processing. In the next example, the view SOFTWARE_PRODUCTS_VIEW is referenced in a SELECT query and arranged in ascending order by product name (PRODNAME).

SQL Code

```
PROC SQL;
  SELECT *
    FROM SOFTWARE_PRODUCTS_VIEW
      ORDER BY prodname;
QUIT;
```

Results

The SAS System

Product Number	Product Name	Manufacturer Number	Product Type
5002	Database Software	500	Software
5004	Graphics Software	500	Software
5001	Spreadsheet Software	500	Software
5003	Wordprocessor Software	500	Software

8.5 Hiding Logic Complexities

Because complex logic constructs such as multi-way table joins, subqueries, or hard-to-understand data relationships may be beyond the skill of other stuff in your area, you may want to build or customize views so that others can access the information easily. The next example illustrates how a complex query containing a two-way join is constructed and saved as a view to simplify its use by other users.

SQL Code

```
PROC SQL;
  CREATE VIEW PROD_MANF_VIEW AS
    SELECT DISTINCT SUM(prodcost) FORMAT=DOLLAR10.2,
           M.manunum,
           M.manuname
      FROM PRODUCTS AS P, MANUFACTURERS AS M
        WHERE P.manunum  = M.manunum AND
              M.manuname = 'KPL Enterprises';
QUIT;
```

SAS Log Results

```
   PROC SQL;
     CREATE VIEW PROD_MANF_VIEW AS
       SELECT DISTINCT SUM(prodcost) FORMAT=DOLLAR10.2,
              M.manunum,
              M.manuname
         FROM PRODUCTS AS P, MANUFACTURERS AS M
           WHERE P.manunum  = M.manunum AND
                 M.manuname = 'KPL Enterprises';
NOTE: SQL view WORK.PROD_MANF_VIEW has been defined.
   QUIT;
NOTE: PROCEDURE SQL used:
      real time           0.00 seconds
```

In the next example, the PROD_MANF_VIEW is simply referenced in a SELECT query. Because the view's SELECT statement references the product cost (PRODCOST) column with a summary function but does not contain a GROUP BY clause, the note "`The query requires remerging summary statistics back with the original data.`" appears in the SAS log below. This situation causes the sum to be calculated and then remerged with each row in the tables being processed.

SQL Code

```
PROC SQL;
  SELECT *
    FROM PROD_MANF_VIEW;
QUIT;
```

SAS Log Results

```
  PROC SQL;
    SELECT *
     FROM PROD_MANF_VIEW;
NOTE: The query requires remerging summary statistics back with the
original data.
   QUIT;
NOTE: PROCEDURE SQL used:
      real time           0.05 seconds
```

Results

```
                    The SAS System

             Manufacturer
                   Number   Manufacturer Name
  ----------------------------------------------------

  $1,296.00           500   KPL Enterprises
```

8.6 Nesting Views

An important feature of views is that they can be based on other views. This is called nesting. One view can access data that comes through another view. In fact there isn't a limit to the number of view layers that can be defined. Because of this, views can be a very convenient and flexible way for programmers to retrieve information. Although the number of views that can be nested is virtually unlimited, programmers should use care to avoid nesting views too deeply. Performance- and maintenance-related issues can result, especially if the views are built many layers deep.

To see how views can be based on other views, two views will be created — one referencing the PRODUCTS table and the other referencing the INVOICE table. In the first example, WORKSTATION_PRODUCTS_VIEW includes only products related to workstations and excludes the manufacturer number. When accessed from the SAS Windowing environment, the view produces the results displayed below.

SQL Code

```
PROC SQL;
  CREATE VIEW WORKSTATION_PRODUCTS_VIEW AS
    SELECT PRODNUM, PRODNAME, PRODTYPE, PRODCOST
      FROM PRODUCTS
        WHERE UPCASE(PRODTYPE)="WORKSTATION";
QUIT;
```

Results

	Product Number	Product Name	Product Type	Product Cost
1	1110	Dream Machine	Workstation	$3,200.00
2	1200	Business Machine	Workstation	$3,300.00

In the next example, INVOICE_1K_VIEW includes rows where the invoice price is $1,000.00 or greater and excludes the manufacturer number. When accessed from the SAS Windowing environment, the view renders the results displayed below.

SQL Code

```
PROC SQL;
  CREATE VIEW INVOICE_1K_VIEW AS
    SELECT INVNUM, CUSTNUM, PRODNUM, INVQTY, INVPRICE
      FROM INVOICE
        WHERE INVPRICE >= 1000.00;
QUIT;
```

Results

	Invoice Number	Customer Number	Product Number	Invoice Quantity - Units Sold	Invoice Unit Price
1	1001	201	5001	5	$1,495.00
2	1002	1301	6001	2	$1,598.00
3	1004	501	1110	3	$9,600.00
4	1007	401	1200	7	$23,100.00

The next example illustrates creating a view from the join of the WORKSTATION_PRODUCTS_VIEW and INVOICE_1K_VIEW views. The resulting view is nested two layers deep. When accessed from the SAS Windowing environment, the view renders the results displayed below.

SQL Code

```
PROC SQL;
  CREATE VIEW JOINED_VIEW AS
    SELECT V1.PRODNUM, V1.PRODNAME,
           V2.CUSTNUM, V2.INVQTY, V2.INVPRICE
      FROM WORKSTATION_PRODUCTS_VIEW  V1,
           INVOICE_1K_VIEW   V2
        WHERE V1.PRODNUM = V2.PRODNUM;
QUIT;
```

Results

	Product Number	Product Name	Customer Number	Invoice Quantity - Units Sold	Invoice Unit Price
1	1110	Dream Machine	501	3	$9,600.00
2	1200	Business Machine	401	7	$23,100.00

In the next example, a third layer of view is nested to the previous view in order to find the largest invoice amount. In the next example, a view is constructed to find the largest invoice amount using the MAX summary function to compute the product of the invoice price (INVPRICE) and invoice quantity (INVQTY) from the JOINED_VIEW view.

When accessed from the SAS Windowing environment, the view produces the results displayed below.

SQL Code

```
PROC SQL;
  CREATE VIEW LARGEST_AMOUNT_VIEW AS
    SELECT MAX(INVPRICE*INVQTY) AS Maximum_Price
           FORMAT=DOLLAR12.2
           LABEL="Largest Invoice Amount"
      FROM JOINED_VIEW;
QUIT;
```

Results

	Largest Invoice Amount
1	$161,700.00

8.7 Updatable Views

Once a view has been created from a physical table, it can then be used to modify the view's data of a single underlying table. Essentially when a view is updated the changes pass through the view to the underlying base table. A view designed in this manner is called an updatable view and can have INSERT, UPDATE, and DELETE operations performed into the single table from which it's constructed.

Because views are dependent on getting their data from a base table and have no physical existence of their own, you should exercise care when constructing an updatable view. Although useful in modifying the rows in a table, updatable views do have a few limitations that programmers and users should be aware of.

First, an updatable view can only have a single base table associated with it. This means that the underlying table cannot be used in a join operation or with any set operators. Because an updatable view has each of its rows associated with just a single row in an underlying table, any operations involving two or more tables will produce an error and result in update operations not being performed.

An updatable view cannot contain a subquery. A subquery is a complex query consisting of a SELECT statement contained inside another statement. This violates the rules for updatable views and is not allowed.

An updatable view can update a column using a view's column alias, but cannot contain the DISTINCT keyword, have any aggregate (summary) functions, calculated columns, or derived columns associated with it. Because these columns are produced by an expression, they are not allowed.

Finally, an updatable view can contain a WHERE clause but not other clauses such as ORDER BY, GROUP BY, or HAVING.

In the remaining sections, three types of updatable views will be examined:

- views that insert one or more rows of data,
- views that update existing rows of data, and
- views that delete one or more rows of data from a single underlying table.

8.7.1 Inserting New Rows of Data

You can add or insert new rows of data in a view using the INSERT INTO statement. Suppose we have a view consisting of only software products called SOFTWARE_PRODUCTS_VIEW. The PROC SQL code used to create this view consists of a SELECT statement with a WHERE clause. There are four defined columns: product name, product number, product type, and product cost, in that order. When accessed from the SAS Windowing environment, the view produces the results displayed below.

SQL Code

```
PROC SQL;
  CREATE VIEW SOFTWARE_PRODUCTS_VIEW AS
    SELECT prodnum, prodname, prodtype, prodcost
           FORMAT=DOLLAR8.2
      FROM PRODUCTS
        WHERE UPCASE(PRODTYPE) IN ('SOFTWARE');
QUIT;
```

Results

	Product Number	Product Name	Product Type	Product Cost
1	5001	Spreadsheet Software	Software	$299.00
2	5002	Database Software	Software	$399.00
3	5003	Wordprocessor Software	Software	$299.00
4	5004	Graphics Software	Software	$299.00

Suppose you want to add a new row of data to this view. This can be accomplished by specifying the corresponding values in a VALUES clause as follows.

SQL Code

```
PROC SQL;
  INSERT INTO SOFTWARE_PRODUCTS_VIEW
    VALUES(6002,'Security Software','Software',375.00);
QUIT;
```

As seen from the view results, the INSERT INTO statement added the new row of data corresponding to the WHERE logic in the view. The view contains the new row and consists of the value 6002 in product number, “Security Software” in product name, “Software” in product type, and $375.00 in product cost, as shown.

View Results

	Product Number	Product Name	Product Type	Product Cost
1	5001	Spreadsheet Software	Software	$299.00
2	5002	Database Software	Software	$399.00
3	5003	Wordprocessor Software	Software	$299.00
4	5004	Graphics Software	Software	$299.00
5	6002	Security Software	Software	$375.00

As depicted in the table results, the new row of data was added to the PRODUCTS table using the view called SOFTWARE_PRODUCTS_VIEW. The new row in the PRODUCTS table contains the value 6002 in product number, “Security Software” in product name, “Software” in product type, and $375.00 in product cost. The manufacturer number column is assigned a null value (missing value), as shown.

Table Results

	Product Number	Product Name	Manufacturer Number	Product Type	Product Cost
1	1110	Dream Machine	111	Workstation	$3,200.00
2	1200	Business Machine	120	Workstation	$3,300.00
3	1700	Travel Laptop	170	Laptop	$3,400.00
4	2101	Analog Cell Phone	210	Phone	$35.00
5	2102	Digital Cell Phone	210	Phone	$175.00
6	2200	Office Phone	220	Phone	$130.00
7	5001	Spreadsheet Software	500	Software	$299.00
8	5002	Database Software	500	Software	$399.00
9	5003	Wordprocessor Software	500	Software	$299.00
10	5004	Graphics Software	500	Software	$299.00
11	6002	Security Software	.	Software	$375.00

Now let’s see what happens when a row of data is added through a view that does not meet the condition(s) in the WHERE clause in the view. Suppose we want to add a row of data containing the value 1701 for product number, “Travel Laptop SE” in product name, “Laptop” in product type, and $4200.00 in product cost in the SOFTWARE_PRODUCTS_VIEW view.

SQL Code

```
PROC SQL;
  INSERT INTO SOFTWARE_PRODUCTS_VIEW
    VALUES(1701,'Travel Laptop SE','Laptop',4200.00);
QUIT;
```

Because the new row’s value for product type is “Laptop”, this value violates the WHERE clause condition when the view SOFTWARE_PRODUCTS_VIEW was created. As a result, the new row of data is rejected and is not added to the table PRODUCTS. The SQL procedure also prevents the new row from appearing in the view because the base table controls what the view contains.

The updatable view does exactly what it is designed to do — that is, validate each new row of data as each row is added to the base table. Whenever the WHERE clause condition is violated the view automatically rejects the row as invalid and restores the table to its pre-updated state by rejecting the row in error and deleting all successful inserts before the error occurred. In our example, we see the following error message was issued to the SAS log to confirm that the view was restored to its original state before the update took place.

SAS Log Results

```
      PROC SQL;
        INSERT INTO PRODUCTS_VIEW
          VALUES(1701,'Travel Laptop SE','Laptop',4200.00);
ERROR: The new values do not satisfy the view's where expression. This update
or add is not allowed.
NOTE: This insert failed while attempting to add data from VALUES clause 1 to
the dataset.
NOTE: Deleting the successful inserts before error noted above to restore table
to a consistent state.
      QUIT;
NOTE: The SAS System stopped processing this step because of errors.
NOTE: PROCEDURE SQL used:
      real time           0.04 seconds
```

Views will not accept new rows added to a base table when the number of columns in the VALUES clause does not match the number of columns defined in the view, unless the columns that are being inserted are specified. In the next example, a partial list of columns for a row of data is inserted with a VALUES clause. Because the inserted row of data does not contain a value for product cost, the new row will not be added to the PRODUCTS table. The resulting error message indicates that the VALUES clause has fewer columns specified than exist in the view itself, as shown in the SAS log below.

SQL Code

```
PROC SQL;
  INSERT INTO SOFTWARE_PRODUCTS_VIEW
   VALUES(6003,'Cleanup Software','Software');
QUIT;
```

SAS Log Results

```
   PROC SQL;
     INSERT INTO SOFTWARE_PRODUCTS_VIEW
      VALUES(6003,'Cleanup Software','Software');
ERROR: VALUES clause 1 attempts to insert fewer columns than specified after
the INSERT table name.
   QUIT;
NOTE: The SAS System stopped processing this step because of errors.
NOTE: PROCEDURE SQL used:
      real time           0.00 seconds
```

Suppose a view called SOFTWARE_PRODUCTS_TAX_VIEW was created with the sole purpose of deriving each software product's sales tax amount as follows.

SQL Code

```
PROC SQL;
  CREATE VIEW SOFTWARE_PRODUCTS_TAX_VIEW AS
    SELECT prodnum, prodname, prodtype, prodcost,
           prodcost * .07 AS Tax
           FORMAT=DOLLAR8.2 LABEL='Sales Tax'
      FROM PRODUCTS
        WHERE UPCASE(PRODTYPE) IN ('SOFTWARE');
QUIT;
```

In the next example, an attempt is made to add a new row through the SOFTWARE_PRODUCTS_TAX_VIEW view by inserting a VALUES clause with all columns defined. The row is rejected and an error produced because an update was attempted against a view that contains a computed (calculated) column. Although the VALUES clause contains values for all columns defined in the view, the reason the row is not inserted into the PRODUCTS table is due to the reference to a computed (or derived) column TAX (Sales Tax) as shown in the SAS log results.

SQL Code

```
PROC SQL;
  INSERT INTO SOFTWARE_PRODUCTS_TAX_VIEW
   VALUES(6003,'Cleanup Software','Software',375.00,26.25);
QUIT;
```

SAS Log Results

```
   PROC SQL;
     INSERT INTO SOFTWARE_PRODUCTS_TAX_VIEW
      VALUES(6003,'Cleanup Software','Software',375.00,26.25);
WARNING: Cannot provide Tax with a value because it references a derived column
that can't be inserted into.
   QUIT;
NOTE: The SAS System stopped processing this step because of errors.
NOTE: PROCEDURE SQL used:
      real time           0.00 seconds
```

8.7.2 Updating Existing Rows of Data

The SQL procedure permits rows to be updated through a view. The data manipulation language statement that is specified to modify existing data in PROC SQL is the UPDATE statement. Suppose a view were created to select only laptops from the PRODUCTS table. The SQL procedure code used to create the view is called LAPTOP_PRODUCTS_VIEW and consists of a SELECT statement with a WHERE clause. There are four defined columns: product name, product number, product type, and product cost, in that specific order. When accessed, the view produces the results displayed below.

SQL Code

```
PROC SQL;
  CREATE VIEW LAPTOP_PRODUCTS_VIEW AS
    SELECT PRODNUM, PRODNAME, PRODTYPE, PRODCOST
      FROM PRODUCTS
        WHERE UPCASE(PRODTYPE) = 'LAPTOP';
QUIT;
```

Results

	Product Number	Product Name	Product Type	Product Cost
1	1700	Travel Laptop	Laptop	$2,720.00

In the next example, all laptops are to be discounted by twenty percent and the new price is to take effect immediately. The changes applied through the LAPTOP_PRODUCTS_VIEW view computes the discounted product cost for “Laptop” computers in the PRODUCTS table using an UPDATE statement with corresponding SET clause.

SQL Code

```
PROC SQL;
  UPDATE LAPTOP_PRODUCTS_VIEW
    SET PRODCOST = PRODCOST - (PRODCOST * 0.2);
QUIT;
```

SAS Log Results

```
      PROC SQL;
         UPDATE LAPTOP_DISCOUNT_VIEW
         SET PRODCOST = PRODCOST - (PRODCOST * 0.2);
NOTE: 1 row was updated in WORK.LAPTOP_DISCOUNT_VIEW.
      QUIT;
NOTE: PROCEDURE SQL used:
      real time             0.04 seconds
```

Results

	Product Number	Product Name	Product Type	Product Cost
1	1700	Travel Laptop	Laptop	$2,720.00

Sometimes updates applied through a view can change the rows of data in the base table so that once the update is performed the rows in the base table no longer meet the criteria in the view. When this occurs, the changed rows of data cannot be displayed by the view. Essentially the updated rows matching the conditions in the WHERE clause no longer match the conditions in the view’s WHERE clause after the updates are made. As a result, the view updates the rows with the specified changes but is no longer able to display the rows of data that were changed.

Suppose a view were created to select laptops costing more than $2,800 from the PRODUCTS table. The SQL procedure code used to create the view called LAPTOP_DISCOUNT_VIEW consists of a SELECT statement with a WHERE clause.

There are four defined columns: product name, product number, product type, and product cost, in that order. When accessed, the view produces the results displayed below.

SQL Code

```
PROC SQL;
  CREATE VIEW LAPTOP_DISCOUNT_VIEW AS
    SELECT PRODNUM, PRODNAME, PRODTYPE, PRODCOST
      FROM PRODUCTS
        WHERE UPCASE(PRODTYPE) = 'LAPTOP' AND
              PRODCOST > 2800.00;
QUIT;
```

Results

	Product Number	Product Name	Product Type	Product Cost
1	1700	Travel Laptop	Laptop	$3,400.00

The next example illustrates how updates are applied through a view in the Windows environment so the rows in the table no longer meet the view's criteria. Suppose a twenty percent discount is applied to all laptops. An UPDATE statement and SET clause are specified to allow the rows in the Products table to be updated through the view. Once the update is performed and the view is accessed, a dialog box appears indicating that no rows are available to display because the data from the PRODUCTS table no longer meets the view's WHERE clause expression.

SQL Code

```
PROC SQL;
  UPDATE LAPTOP_DISCOUNT_VIEW
    SET PRODCOST = PRODCOST - (PRODCOST * 0.2);
QUIT;
```

SAS Log Results

```
     PROC SQL;
       UPDATE LAPTOP_DISCOUNT_VIEW
         SET PRODCOST = PRODCOST - (PRODCOST * 0.2);
NOTE: 1 row was updated in WORK.LAPTOP_DISCOUNT_VIEW.
     QUIT;
NOTE: PROCEDURE SQL used:
     real time           0.06 seconds
```

Results

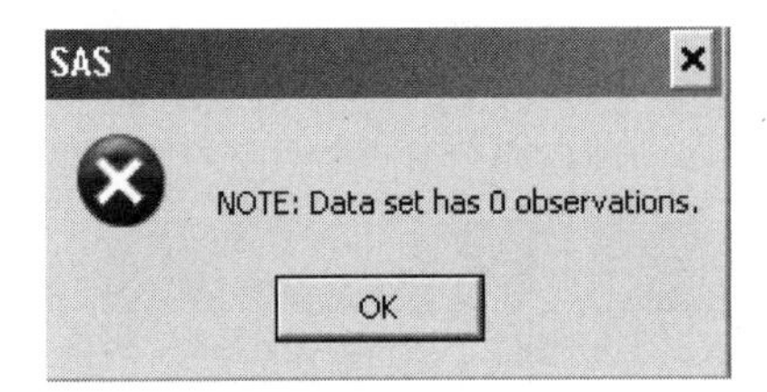

8.7.3 Deleting Rows of Data

Now that you have seen how updatable views can add or modify one or more rows of data, you may have a pretty good idea how to create an updatable view that deletes one or more rows of data. Consider the following updatable view that deletes manufacturers whose manufacturer number is 600 from the underlying PRODUCTS table.

SQL Code

```
PROC SQL;
  DELETE FROM SOFTWARE_PRODUCTS_VIEW
    WHERE MANUNUM=600;
QUIT;
```

SAS Log Results

```
   PROC SQL;
     DELETE FROM SOFTWARE_PRODUCTS_VIEW
       WHERE MANUNUM=600;
NOTE: 2 rows were deleted from WORK.SOFTWARE_PRODUCTS_VIEW.
   QUIT;
NOTE: PROCEDURE SQL used:
      real time           0.04 seconds
```

8.8 Deleting Views

When a view is no longer needed, it's nice to know there is a way to remove it. Without this ability, program maintenance activities would be made more difficult. To remove an unwanted view, specify the DROP VIEW statement and the name of the view. In the next example, the INVENTORY_VIEW view is deleted from the WORK library.

SQL Code

```
PROC SQL;
  DROP VIEW INVENTORY_VIEW;
QUIT;
```

SAS Log

```
     PROC SQL;
       DROP VIEW INVENTORY_VIEW;
NOTE: View WORK.INVENTORY_VIEW has been dropped.
     QUIT;
NOTE: PROCEDURE SQL used:
      real time           0.10 seconds
```

When more than a single view needs to be deleted, the DROP VIEW statement works equally as well. Specify a comma between each view name when deleting two or more views.

SQL Code

```
PROC SQL;
  DROP VIEW INVENTORY_VIEW, LAPTOP_PRODUCTS_VIEW;
QUIT;
```

SAS Log

```
      PROC SQL;
        DROP VIEW INVENTORY_VIEW, LAPTOP_PRODUCTS_VIEW;
NOTE: View WORK.INVENTORY_VIEW has been dropped.
NOTE: View WORK.LAPTOP_PRODUCTS_VIEW has been dropped.
      QUIT;
NOTE: PROCEDURE SQL used:
       real time           0.00 seconds
```

8.9 Summary

1. Views are not tables and consequently do not store data (see section 8.2).
2. Views access one or more underlying tables (base tables) or other views (see section 8.2).
3. Views improve the change control process when constructed as a "common" set of routines (see section 8.2).
4. Views eliminate or reduce data redundancy (see section 8.3).
5. Views hide unwanted or sensitive information while displaying specific columns and/or rows (see section 8.4).
6. Views shield users from making logic and/or data errors (see section 8.5).
7. Nesting views too deeply can produce unnecessary confusion and maintenance difficulties (see section 8.6).

8. Updatable views add, modify, or delete rows of data (see section 8.7).

9. Views can be deleted when no longer needed (see section 8.8).

Chapter 9

Troubleshooting and Debugging

9.1 Introduction

When it comes to tracking down the source of SQL coding problems, users can be very resourceful. To find all those pesky errors, warnings, notes, and unexpected results, you need to carefully inspect the SAS log and telltale clues from code reviews and sample runs. Using reliable troubleshooting and debugging techniques will help get to the root of any syntax, data, system-related, and logic errors that are or could be problematic to the successful execution of a PROC SQL program.

This chapter introduces a number of strategies and techniques for troubleshooting and debugging SQL procedure coding problems. The guidelines presented in this chapter will help you learn about the different types of bugs, the steps involved in troubleshooting problems, the types of errors that can crop up, and the options and statements that are available in the SQL procedure to troubleshoot and debug problems.

9.2 The World of Bugs

Ever since the birth of the software industry, software problems (often referred to as bugs) have been created by programmers (and users) of all skill levels in virtually every conceivable form. A bug is something the software or a program does that it is not supposed to do. When the software encounters a bug, it can cause the software to cease to operate or misbehave.

Bugs come in all forms, shapes, and sizes. A problem that occurs while you're using the SQL procedure incorrectly and directly violating the rules of the language is referred to as a usage error. The severity of these types of errors depends on the nature of the violation as well as the effect on the user. Generally speaking though, a usage error results in a syntax error and the stoppage of the program or step.

Other types of bugs can be problematic too. Bugs such as resource problems, data dependencies, and implementation errors can cause a program step to stop, or produce unreliable results, displaying errors or warnings in the SAS log.

9.3 The Debugging Process

The debugging process consists of a number of recommended steps to correct an identified problem and verify that it does not reappear. The objective is to identify, classify, fix, and verify a problem in a PROC SQL step, program, or application as quickly and easily as possible. Not every problem requires the application of each step recommended in this chapter. Rather, you can pick and choose from the suggestions to effectively correct problems and make sure they do not reappear. Adhering to a methodical and effective debugging process increases the likelihood that problems are identified and fixed correctly, thereby expediting and improving the way you handle problems.

To further explore this topic see *Effective Methods for Software Testing* by William Perry (John Wiley & Sons, Inc., 1995) and *The Science of Debugging* by Matt Telles and Yuan Hsieh (The Coriolis Group, 2001).

The debugging process consists of five steps.

Debugging Process Steps and Tasks

Debugging Step	Task Description
Problem Identification	1. Determine if a bug exists in the code. 2. Determine what the problem is by playing detective. 3. Describe why the problem is a bug. 4. Determine what the code should do. 5. Determine what the code is doing.
Information Collection	1. Collect user comments and feedback. 2. Collect personal observations and symptoms. 3. Review SAS log information. 4. Collect test case(s) highlighting the problem. 5. Capture environmental information (for example, system settings, operating environment, external elements, etc.).
Problem Assessment & Classification	1. Develop a theory about what caused the problem. 2. Review the code. 3. Classify the problem into one of the following categories: - Requirements problem - Syntax error - CPU problem - Memory problem - Storage problem - I/O problem - Logic problem

(continued on next page)

Debugging Step	Task Description
Problem Resolution	1. Propose a solution. 2. Describe why the solution will fix the problem. 3. Verify that the solution will not cause additional problems. 4. Fix the problem by implementing the solution.
Validate Solution	1. Verify whether the problem still exists. 2. Verify whether the problem can be recreated or reproduced. 3. Determine whether other methods can cause the same problem to occur. 4. Verify that the solution does not cause other problems.

9.4 Types of Problems

The leading causes of SQL programming problems include misusing the language syntax; referencing data, especially column data, incorrectly; ignoring or incorrectly specifying system parameters, and constructing syntactically correct but illogical code that does not produce results as expected.

Usage errors can cause the SAS System to stop processing, produce warnings, or produce unexpected results. The following table illustrates the four types of usage errors and briefly describes each.

Types of Usage Errors

Problem	Description
Syntax	Syntax problems are a result of one or more violations of the SQL procedure language constructs. These problems can prevent a program from processing until you make the required changes. For example, a variable that is referenced in a SELECT statement but not found in the referenced table causes the program to stop.
Data	Data problems are a result of an inconsistency between the data and the program specification. These problems can prevent a program from processing, but may allow processing to continue resulting in the assignment of missing, incomplete, or unreliable data. For example, missing values may be generated in a column when incorrectly referencing character data as numeric data.
System-related	System-related problems frequently result from specifying incompatible system options or choosing the wrong system option values. These problems can prevent a program from processing, but most frequently permit processing to occur with unsatisfactory results. For example, forgetting to specify a title statement has little effect on the production of output but may force the program to be rerun after the addition of one or more titles.
Logic	Logic problems are frequently the result of not specifying a coding condition correctly. For example, specifying an OR condition when an AND condition was needed may produce wrong results without any warnings, errors, or notes.

Other sources of problems can creep into a program for a variety of reasons. The following table illustrates these other problem sources and briefly describes each.

Other Types of Errors

Problem	Description
Feature Creep	Additional features can creep into a program or application during the design, implementation, or testing phases. This can create a greater likelihood of problems.
Solution Complexity	More complex or esoteric solutions can also translate into problems. Difficult-to-maintain programs can result.
Requirements	Requirements may be inadequately stated, misunderstood, or omitted. In these situations, programs may not meet the needs of the user community and be classified as bugs.
Testing Environment	Inadequate testing environments, poor test plans, or insufficient test time often lead to bug-filled programs. In these situations, bugs can slip through the testing phase and into the production environment.

9.5 Troubleshooting and Debugging Techniques

PROC SQL provides numerous troubleshooting and debugging capabilities for the practitioner to choose from. From statements to options to macro variables, you can use the various techniques to control the process of finding and gathering information about SQL procedure coding problems quickly and easily.

9.5.1 Validating Queries with the VALIDATE Statement

The SQL procedure syntax checker identifies syntax errors before any data is processed. Syntax checking is automatically turned on in the SQL procedure without any statements or options being specified. But to enable syntax checking without automatically executing a step, you can specify a VALIDATE statement at the beginning of a SELECT statement or, as will be presented later, specify the NOEXEC option.

The VALIDATE statement is available for SAS users to control what SELECT statement is checked. Because you specify it before a SELECT statement, you are better able to control the process of debugging code. A message indicating syntax correctness is automatically displayed in the SAS log when code syntax is valid. Otherwise, an error message is displayed identifying the coding violation. In the next example, a VALIDATE statement is specified at the beginning of a SELECT statement to enable syntax checking without code execution. The SAS log shows that the code contains valid syntax and automatically displays a message to that effect.

SQL Code

```
PROC SQL;
 VALIDATE
  SELECT *
   FROM PRODUCTS
    WHERE PRODTYPE = 'Software';
QUIT;
```

SAS Log Results

```
PROC SQL;
 VALIDATE
  SELECT *
   FROM PRODUCTS
    WHERE PRODTYPE = 'Software';
NOTE: PROC SQL statement has valid syntax.
QUIT;
```

9.5.2 Documented PROC SQL Options and Statement

This section shows several examples of widely used, documented SQL options and the RESET statement. A description along with a working example of each option and statement is presented below.

9.5.2.1 FEEDBACK Option

The FEEDBACK option displays additional documentation with SELECT queries. When specified, this option expands a SELECT * (wildcard) statement into a list of columns that it represents by displaying the names of each column in the underlying table(s) as well as any resolved macro values and macro variables. The column display order of a SELECT * statement is determined by the order defined in the table's record descriptor.

The following example illustrates the FEEDBACK option. Because a SELECT * statement does not automatically display the columns it represents, it may be important to expand the individual column names by specifying the FEEDBACK option. This becomes particularly useful for determining whether a desired column is present in the output and available for documentation purposes. The results of the expanded list of columns are displayed in the SAS log.

SQL Code

```
PROC SQL FEEDBACK;
 SELECT *
  FROM PRODUCTS;
QUIT;
```

SAS Log Results

```
PROC SQL FEEDBACK;
  SELECT *
    FROM PRODUCTS;

NOTE: Statement transforms to:

  select PRODUCTS.prodnum, PRODUCTS.prodname, PRODUCTS.manunum,
         PRODUCTS.prodtype, PRODUCTS.prodcost
       from WORK.PRODUCTS;
QUIT;
```

The FEEDBACK option can be particularly helpful in determining the column order when joining two or more tables. The next example illustrates the expansion of the columns in the SELECT * statement in a two-way equijoin. The FEEDBACK option displays all the columns in both tables.

SQL Code

```
PROC SQL FEEDBACK;
   SELECT *
      FROM PRODUCTS, MANUFACTURERS
        WHERE PRODUCTS.MANUNUM = MANUFACTURERS.MANUNUM AND
              MANUFACTURERS.MANUNAME = 'KPL Enterprises';
QUIT;
```

SAS Log Results

```
PROC SQL FEEDBACK;
  SELECT *
    FROM PRODUCTS, MANUFACTURERS
      WHERE PRODUCTS.MANUNUM = MANUFACTURERS.MANUNUM AND
            MANUFACTURERS.MANUNAME = 'KPL Enterprises';

NOTE: Statement transforms to:

  select PRODUCTS.prodnum, PRODUCTS.prodname, PRODUCTS.manunum,
         PRODUCTS.prodtype, PRODUCTS.prodcost, MANUFACTURERS.manunum,
         MANUFACTURERS.manuname, MANUFACTURERS.manucity,
         MANUFACTURERS.manustat
    from PRODUCTS, MANUFACTURERS
      where (PRODUCTS.manunum=MANUFACTURERS.manunum) and
            (MANUFACTURERS.manuname='KPL Enterprises');

QUIT;
```

The FEEDBACK option can also be used to display macro value and macro variable resolution. The next example shows the macro resolution of the macro variables &LIB, &TABLE, and &GROUPBY for debugging purposes.

SQL Code

```
%MACRO DUPS(LIB, TABLE, GROUPBY);
  PROC SQL FEEDBACK;
    SELECT &GROUPBY, COUNT(*) AS Duplicate_Rows
      FROM &LIB..&TABLE
        GROUP BY &GROUPBY
          HAVING COUNT(*) > 1;
  QUIT;
%MEND DUPS;

%DUPS(WORK,PRODUCTS,PRODTYPE);
```

SAS Log Results

```
%MACRO DUPS(LIB, TABLE, GROUPBY);
  PROC SQL FEEDBACK;
    SELECT &GROUPBY, COUNT(*) AS Duplicate_Rows
      FROM &LIB..&TABLE
        GROUP BY &GROUPBY
          HAVING COUNT(*) > 1;
  QUIT;
%MEND DUPS;

%DUPS(WORK,PRODUCTS,PRODTYPE);

NOTE: Statement transforms to:

    select PRODUCTS.prodtype, COUNT(*) as Duplicate_Rows
      from WORK.PRODUCTS
        group by PRODUCTS.prodtype
          having COUNT(*)>1;
```

You can also specify a %PUT statement instead of the FEEDBACK option to display the values of macro variables after macro resolution. The next example illustrates inserting the macro statement %PUT LIB = &LIB TABLE = &TABLE GROUPBY = &GROUPBY between the QUIT and %MEND statements to produce the results illustrated below.

SQL Code

```
%MACRO DUPS(LIB, TABLE, GROUPBY);
  PROC SQL;
    SELECT &GROUPBY, COUNT(*) AS Duplicate_Rows
      FROM &LIB..&TABLE
        GROUP BY &GROUPBY
          HAVING COUNT(*) > 1;
  QUIT;
  %PUT LIB = &LIB TABLE = &TABLE GROUPBY = &GROUPBY;
%MEND DUPS;

%DUPS(WORK,PRODUCTS,PRODTYPE);
```

SAS Log Results

```
%MACRO DUPS(LIB, TABLE, GROUPBY);

. . . code not shown . . .

LIB = WORK TABLE = PRODUCTS GROUPBY = PRODTYPE
```

9.5.2.2 INOBS= Option

The INOBS= option reduces the amount of query execution time by restricting the number of rows that PROC SQL processes. This option is most often used for troubleshooting or debugging purposes where a small number of rows are needed as opposed to all the rows in the table source. Controlling the number of rows processed on input with the INOBS= option is similar to specifying the SAS System option to OBS=. The following example illustrates the INOBS= option being limited to the first ten rows in the PRODUCTS table. A warning message is also displayed in the SAS log indicating that the number of records read was restricted to ten.

SQL Code

```
PROC SQL INOBS=10;
 SELECT *
  FROM PRODUCTS;
QUIT;
```

SAS Log Results

```
PROC SQL INOBS=10;
  SELECT *
    FROM PRODUCTS;
WARNING: Only 10 records were read from WORK.PRODUCTS due to INOBS= option.
QUIT;
```

In the next example, the INOBS= option is set to five in a Cartesian product join (the absence of a WHERE clause). A two-way join with the INOBS=5 specified without a WHERE clause limits the number of rows from each table to five producing a maximum of 25 rows.

SQL Code

```
PROC SQL INOBS=5;
 SELECT prodname, prodcost,
        manufacturers.manunum, manuname
    FROM PRODUCTS, MANUFACTURERS;
QUIT;
```

SAS Log Results

```
PROC SQL INOBS=5;
  SELECT prodname, prodcost,
         manufacturers.manunum, manuname
    FROM PRODUCTS, MANUFACTURERS;
NOTE: The execution of this query involves performing one or more Cartesian
product joins that can not be optimized.
WARNING: Only 5 records were read from WORK.MANUFACTURERS due to INOBS= option.
WARNING: Only 5 records were read from WORK.PRODUCTS due to INOBS= option.
QUIT;
```

9.5.2.3 LOOPS= Option

The LOOPS= option reduces the amount of query execution time by restricting how many times processing occurs through a query's inner loop. As with the INOBS= and OUTOBS= options, the LOOPS= option is used for troubleshooting or debugging to prevent the consumption of excess computer resources or the creation of large internal tables as with the processing of multi-table joins. The following example shows the LOOPS= option being restricted to eight inner loops through the rows in the PRODUCTS table.

SQL Code

```
PROC SQL LOOPS=8;
 SELECT *
  FROM PRODUCTS;
QUIT;
```

SAS Log Results

```
PROC SQL LOOPS=8;
  SELECT *
    FROM PRODUCTS;
WARNING: PROC SQL statement interrupted by LOOPS=8 option.
QUIT;
```

Results

```
                                    The SAS System

 Product                              Manufacturer                        Product
  Number  Product Name                      Number  Product Type             Cost
 ________________________________________________________________________________

    1110  Dream Machine                        111  Workstation         $3,200.00
    1200  Business Machine                     120  Workstation         $3,300.00
    1700  Travel Laptop                        170  Laptop              $3,400.00
```

The next example shows what happens when the LOOPS= option is applied in a three-way join by restricting the number of processed inner loops to 50 to prevent the creation of a large and inefficient internal table. To determine an adequate value to assign to the LOOPS= option, you can specify an &SQLOOPS macro variable in a %PUT statement. To learn more about this macro variable, see the section, "Macro Variables," later in this chapter.

SQL Code

```
PROC SQL LOOPS=50;
 SELECT P.prodname, P.prodcost,
        M.manuname,
        I.invqty
   FROM PRODUCTS  P,
        MANUFACTURERS  M,
        INVOICE  I
     WHERE P.manunum = M.manunum AND
           P.prodnum = I.prodnum AND
           M.manunum = 500;
QUIT;
```

Results

The SAS System

Product Name	Product Cost	Manufacturer Name	Invoice Quantity - Units Sold
Spreadsheet Software	$299.00	KPL Enterprises	5
Database Software	$399.00	KPL Enterprises	2

9.5.2.4 NOEXEC Option

The NOEXEC option checks all nonquery statements such as CREATE TABLE or ALTER TABLE within the SQL procedure for syntax-related errors and displays any identified errors in the SAS log. The NOEXEC option is similar to the VALIDATE statement because both of them check for syntax correctness without the execution of any input data. The only difference between the NOEXEC option and the VALIDATE statement is in the way each is specified. As was presented earlier, the VALIDATE statement is specified before each SELECT statement; the NOEXEC option is specified only once as an option in the PROC SQL statement. The NOEXEC option checks any nonquery statements in the step for syntax correctness. The following example illustrates the NOEXEC option and what happens when an error is found in any nonquery statement.

SQL Code

```
PROC SQL NOEXEC;
  CREATE TABLE NOEXEC_CHECK
    SELECT *
      FROM PRODUCTS
        WHERE PRODTYPE = 'Software';
QUIT;
```

SAS Log Results

```
PROC SQL NOEXEC;
  CREATE TABLE NOEXEC_CHECK
    SELECT *
            ------
            73
 ERROR 73-322: Expecting an AS.
      FROM PRODUCTS
        WHERE PRODTYPE = 'Software';
QUIT;
```

9.5.2.5 OUTOBS= Option

The OUTOBS= option reduces the amount of query execution time by restricting the number of rows that PROC SQL sends as output to a designated output source. As with the INOBS= option, the OUTOBS= option is most often used for troubleshooting or debugging purposes where a small number of rows are needed in an output table. Controlling the number of rows sent as output with the OUTOBS= option is similar to setting the SAS System option OBS= (or the data set option OBS=). The following example creates an output table called PRODUCTS_SAMPLE by specifying five rows for the OUTOBS= option. A warning message displayed in the SAS log indicates that the new table contains five rows.

SQL Code

```
PROC SQL OUTOBS=5;
  CREATE TABLE PRODUCTS_SAMPLE AS
    SELECT *
      FROM PRODUCTS;
QUIT;
```

SAS Log Results

```
PROC SQL OUTOBS=5;
  CREATE TABLE PRODUCTS_SAMPLE AS
    SELECT *
      FROM PRODUCTS;
WARNING: Statement terminated early due to OUTOBS=5 option.
NOTE: Table WORK.PRODUCTS_SAMPLE created, with 5 rows and 5 columns.
QUIT;
```

9.5.2.6 PROMPT Option

The PROMPT option is issued during interactive sessions to prompt users to continue or stop processing when the limits of an INOBS=, LOOPS=, and/or OUTOBS= option are reached. If the PROMPT option is specified along with one or more of these options, a dialog box appears indicating that the limits of the specified option have been reached and asking whether to stop or continue processing. This prompting feature is a useful process for stepping through a running application.

The following example shows the PROMPT option being issued to initiate a dialogue between the SQL procedure session and the user. The PROMPT option specifies that the INOBS= option limit input processing to the first five rows in the PRODUCTS table. A dialog box automatically appears after five rows are read asking whether processing is to stop or continue. If processing is continued, another five rows are processed and, if additional rows are available for processing, another prompt dialog box appears. This dialogue process continues until all rows are processed or until the user halts processing.

SQL Code

```
PROC SQL PROMPT INOBS=5;
 SELECT *
  FROM PRODUCTS;
QUIT;
```

SAS Log Results

```
PROC SQL PROMPT INOBS=5;
  SELECT *
    FROM PRODUCTS;
WARNING: Only 5 records were read from WORK.PRODUCTS due to INOBS= option.
QUIT;
```

9.5.2.7 RESET Statement

The RESET statement is used to add, drop, or change one or more PROC SQL options without the need of restarting the procedure. Once an option is specified, it stays in effect until it is changed or reset. Being able to change options with the RESET statement is a handy debugging technique. The following example illustrates turning off the FEEDBACK option by resetting it to NOFEEDBACK. By turning off this option, you prevent the expansion of a SELECT * (wildcard) statement into a list of columns that it represents.

SQL Code

```
PROC SQL FEEDBACK;
 SELECT *
  FROM PRODUCTS;

 RESET NOFEEDBACK;

  SELECT *
   FROM PRODUCTS
    WHERE PRODTYPE='Software';
QUIT;
```

SAS Log Results

```
SELECT *
 FROM PRODUCTS;
NOTE: Statement transforms to:

select PRODUCTS.prodnum, PRODUCTS.prodname, PRODUCTS.manunum,
       PRODUCTS.prodtype, PRODUCTS.prodcost
 from PRODUCTS;

RESET NOFEEDBACK;

 SELECT *
  FROM PRODUCTS
   WHERE PRODTYPE='Software';
```

Multiple options can be reset in a single RESET statement. Options in the PROC SQL and RESET statements can be specified in any order. The next example shows how, in a single RESET statement, double-spaced output is changed to single-spaced output with the NODOUBLE option, row numbers are suppressed with the NONUMBER option, and output rows are changed to the maximum number of rows with the OUTOBS= option.

SQL Code

```
PROC SQL DOUBLE NUMBER OUTOBS=1;
 SELECT *
  FROM PRODUCTS
   WHERE PRODTYPE='Software';

 RESET NODOUBLE NONUMBER OUTOBS=MAX;

  SELECT *
   FROM PRODUCTS
    WHERE PRODTYPE='Software';
QUIT;
```

SAS Log Results

```
PROC SQL DOUBLE NUMBER OUTOBS=1;
 SELECT *
  FROM PRODUCTS
   WHERE PRODTYPE='Software';
WARNING: Statement terminated early due to OUTOBS=1 option.

 RESET NODOUBLE NONUMBER OUTOBS=MAX;

 SELECT *
  FROM PRODUCTS
   WHERE PRODTYPE='Software';
QUIT;
```

Output Results

```
                              The SAS System

         Product                                Manufacturer                 Product
  Row     Number  Product Name                        Number  Product Type      Cost
 ---------------------------------------------------------------------------------------
    1       5001  Spreadsheet Software                   500  Software       $299.00


                              The SAS System

  Product                           Manufacturer                             Product
   Number  Product Name                   Number  Product Type                  Cost
 ---------------------------------------------------------------------------------
     5001  Spreadsheet Software              500  Software                   $299.00
     5002  Database Software                 500  Software                   $399.00
     5003  Wordprocessor Software            500  Software                   $299.00
     5004  Graphics Software                 500  Software                   $299.00
```

The next example shows a RESET statement being issued to change the way the SQL procedure handles updating data in a table. The first UPDATE query is set to reverse any updates that have been performed up to the point of an error using the UNDO_POLICY=REQUIRED (default value) option. (Note: Because this is the default value for this option, it could have been omitted.) A RESET statement of UNDO_POLICY=NONE is issued before the second update query to change the way updates are handled in PROC SQL. The NONE option keeps any updates that have been made regardless of whether an error is detected. A warning message is displayed on the SAS log alerting you to the change in the way updates are handled.

SQL Code

```
PROC SQL UNDO_POLICY=REQUIRED;
 UPDATE PRODUCTS
  SET PRODCOST = PRODCOST - (PRODCOST * 0.2)
   WHERE UPCASE(PRODTYPE) = 'LAPTOP';

 RESET UNDO_POLICY=NONE;

 UPDATE PRODUCTS
  SET PRODCOST = PRODCOST - (PRODCOST * 0.2)
   WHERE UPCASE(PRODTYPE) = 'LAPTOP';
QUIT;
```

SAS Log Results

```
PROC SQL UNDO_POLICY=REQUIRED;
 UPDATE PRODUCTS
  SET PRODCOST = PRODCOST - (PRODCOST * 0.2)
   WHERE UPCASE(PRODTYPE) = 'LAPTOP';
NOTE: 1 row was updated in WORK.PRODUCTS.

 RESET UNDO_POLICY=NONE;

 UPDATE PRODUCTS
  SET PRODCOST = PRODCOST - (PRODCOST * 0.2)
   WHERE UPCASE(PRODTYPE) = 'LAPTOP';
WARNING: The SQL option UNDO_POLICY=REQUIRED is not in effect. If an
error is detected when processing this UPDATE statement, that error
will not cause the entire statement to fail.
NOTE: 1 row was updated in WORK.PRODUCTS.

QUIT;
```

9.6 Undocumented PROC SQL Options

This section lists several undocumented PROC SQL options. Although undocumented options may be freely explored and used, you should carefully consider the ramifications before using them. Because of unannounced changes, possible removal, or nonsupport in future releases, you should exercise care when using them throughout SQL procedure applications. However, undocumented options provide a wealth of opportunities for identifying and resolving coding problems. The table below presents several of these undocumented SQL procedure options for troubleshooting and debugging purposes.

Undocumented PROC SQL Options

Option	Description
_AGGR	displays a tree structure in the SAS log with a before-and-after summary.
_ASGN	displays a tree structure in the SAS log consisting of resolved before-and-after names.
_DFR	displays a before-and-after dataflow and subcall resolution in a tagged tree structure in the SAS log.
_METHOD	displays the various PROC SQL execution options in the SAS log. Note: This option is explained in greater detail in Chapter 10, "Tuning for Performance and Efficiency."
_PJD	displays various table attributes including the number of observations (rows), the logical record length (lrecl), the number of restricted rows, and the size of the table in bytes.
_RSLV	displays a tree structure in the SAS log consisting of before-and-after early semantic checks.
_SUBQ	displays subquery transformations as a tree structure in the SAS log.
_TREE	displays a query as a tree structure in the SAS log. The tree structure consists of the transformed code that the SQL processor will execute.
_UTIL	displays a breakdown in the SAS log of each step defined in the procedure including each row's buffer length, each column by name, and the column position with each row (or offset).

One of my favorite undocumented PROC SQL options is _TREE. As its name implies, it displays a tree structure or hierarchy illustrating an expanded view of each statement, clause, and column name within the procedure step. This provides a handy and unique way of seeing SQL procedure code to aid in troubleshooting logic and column lists. The next example shows the _TREE option being used in a simple query that specifies the wildcard character "*" that represents all columns with a WHERE clause to subset only "Software" products.

SQL Code

```
PROC SQL _TREE;
  SELECT *
    FROM PRODUCTS
      WHERE PRODTYPE = 'Software';
QUIT;
```

SAS Log Results

```
   PROC SQL _TREE;
     SELECT *
       FROM PRODUCTS
         WHERE PRODTYPE = 'Software';

Tree as planned.
                                      /-SYM-V-(PRODUCTS.prodnum:1 flag=0001)
                           /-OBJ----|
                           |          |--SYM-V-(PRODUCTS.prodname:2 flag=0001)
                           |          |--SYM-V-(PRODUCTS.manunum:3 flag=0001)
                           |          |--SYM-V-(PRODUCTS.prodtype:4 flag=0001)
                           |          \-SYM-V-(PRODUCTS.prodcost:5 flag=0001)
               /-SRC----|
               |           |--TABL[WORK].PRODUCTS opt=''
               |           |          /-NAME--(prodtype:4)
               |           \-CEQ----|
               |                      \-LITC('Software')
 --SSEL---|

   QUIT;
NOTE: PROCEDURE SQL used (Total process time):
      real time           0.19 seconds
      cpu time            0.01 seconds
```

9.6.1 Macro Variables

To assist with the process of troubleshooting and debugging problematic coding constructs, PROC SQL assigns values to three automatic macro variables after the execution of each statement. The contents of these three macro variables can be used to test the validity of SQL procedure code as well as to evaluate whether processing should continue.

9.6.1.2 SQLOBS Macro Variable

The SQLOBS macro variable displays the number of rows that are processed by an SQL procedure statement. To display the contents of the SQLOBS macro variable in the SAS log, specify a %PUT macro statement. The following example retrieves the software products from the PRODUCTS table and displays SAS output with a SELECT statement. The %PUT statement displays the number of rows processed and sent to SAS output as four rows.

SQL Code

```
PROC SQL;
 SELECT *
  FROM PRODUCTS
   WHERE PRODTYPE='Software';

 %PUT SQLOBS = &SQLOBS;

QUIT;
```

SAS Log Results

```
PROC SQL;
 SELECT *
  FROM PRODUCTS
   WHERE PRODTYPE='Software';
 %PUT SQLOBS = &SQLOBS;

 SQLOBS = 4

QUIT;
```

The next example shows two new products inserted in the PRODUCTS table with an INSERT INTO statement. The %PUT statement displays the number of rows added to the PRODUCTS table as two rows.

SQL Code

```
PROC SQL;
 INSERT INTO PRODUCTS
          (PRODNUM, PRODNAME, PRODTYPE, PRODCOST)
  VALUES(6002,'Security Software','Software',375.00)
  VALUES(1701,'Travel Laptop SE', 'Laptop',  4200.00);

 %PUT SQLOBS = &SQLOBS;

QUIT;
```

SAS Log Results

```
PROC SQL;
 INSERT INTO PRODUCTS
          (PRODNUM, PRODNAME, PRODTYPE, PRODCOST)
  VALUES(6002,'Security Software','Software',375.00)
  VALUES(1701,'Travel Laptop SE', 'Laptop',  4200.00);

NOTE: 2 rows were inserted into WORK.PRODUCTS.

 %PUT SQLOBS = &SQLOBS;
 SQLOBS = 2

QUIT;
```

9.6.1.3 SQLOOPS Macro Variable

The SQLOOPS macro variable displays the number of times the inner loop is processed by the SQL procedure. To display the contents of the SQLOOPS macro variable on the SAS log, specify a %PUT macro statement. The following example retrieves the software products from the PRODUCTS table and displays SAS output with a SELECT statement. The %PUT statement displays the number of times the inner loop is processed as 15 times even though there are only 10 product rows. As a query becomes more complex, the number of times the inner loop of the SQL procedure processes also increases proportionally.

SQL Code

```
PROC SQL;
 SELECT *
  FROM PRODUCTS
   WHERE PRODTYPE='Software';

 %PUT SQLOOPS = &SQLOOPS;

QUIT;
```

SAS Log Results

```
PROC SQL;
 SELECT *
  FROM PRODUCTS
   WHERE PRODTYPE='Software';

 %PUT SQLOOPS = &SQLOOPS;
 SQLOOPS = 15

QUIT;
```

9.6.1.4 SQLRC Macro Variable

The SQLRC macro variable displays a status value indicating whether the PROC SQL statement was successful or not. A %PUT macro statement is specified to display the contents of the SQLRC macro variable. The following example retrieves the software products from the PRODUCTS table and displays SAS output with a SELECT statement. The %PUT statement displays a return code of zero indicating that the SELECT statement was successful.

SQL Code

```
PROC SQL;
 SELECT *
  FROM PRODUCTS
   WHERE PRODTYPE='Software';

 %PUT SQLRC = &SQLRC;

QUIT;
```

SAS Log Results

```
PROC SQL;
 SELECT *
  FROM PRODUCTS
   WHERE PRODTYPE='Software';

 %PUT SQLRC = &SQLRC;
 SQLRC = 0

QUIT;
```

9.6.2 Troubleshooting and Debugging Examples

This section shows a number of errors that I have personally experienced while working on SQL procedure problems. Although not representative of all the possible errors that may occur, it does illustrate a set of common problems along with a technical approach for correcting each problem.

ERROR 78-322: Expecting a ','

Problem Description

Syntax errors messages can, at times, provide confusing information about the specific problem at hand. A case in point is the error, **78-322: Expecting a ','**. In the example below, it initially appears that a comma is missing between two column names in the SELECT statement. On closer review, the actual problem points to a violation of the column's naming conventions caused by specifying an invalid character in the assigned column alias in the AS keyword.

Code and Error

```
PROC SQL;
  SELECT CUSTNUM, ITEM, UNITS * UNITCOST AS Total-Cost
                                                     -
                                                     78
ERROR 78-322: Expecting a ','.

    FROM PURCHASES
      ORDER BY TOTAL;
QUIT;
```

Corrective Action

Correct the problem associated with the assigned column-alias name by adhering to valid SAS naming conventions. For example, replace the hyphen "-" in Total-Cost with an underscore, as in Total_Cost.

ERROR 202-322: The option or parameter is not recognized and will be ignored

Problem Description

Sometimes problems occur because of unfamiliarity with the SQL procedure language syntax. In the syntax error illustrated below an unrecognized option or parameter is encountered, resulting in the procedure stopping before any processing occurs.

Code and Error

```
PROC SQL;
  SELECT prodtype,
         MIN(prodcost) AS Cheapest
         Format=dollar9.2 Label='Least Expensive'
    FROM PRODUCTS
      ORDER BY cheapest
        GROUP BY prodtype;
        ----- --
        22    202
ERROR 22-322: Syntax error, expecting one of the following: ;,
!, !!, &, (, *, **, +, ',', -, '.', /, <, <=, <>, =, >, >=, ?,
AND, ASC, ASCENDING, BETWEEN, CONTAINS, DESC, DESCENDING, EQ,
EQT, GE, GET, GT, GTT, IN, IS, LE, LET, LIKE, LT, LTT, NE, NET,
NOT, NOTIN, OR, ^, ^=, |, ||, ~, ~=.
```

```
ERROR 202-322: The option or parameter is not recognized and
will be ignored.

QUIT;
```

Corrective Action

This problem is the result of the SELECT statement clauses not being specified in the correct order. It can be corrected by specifying the SELECT statement's GROUP BY clause before the ORDER BY clause.

ERROR Ambiguous reference, column

Problem Description

In the next example, the syntax error points to a problem where a column name that is specified in a SELECT statement appears in more than one table resulting in a column ambiguity. This problem not only creates confusion for the SQL processor, but also prevents the query from executing.

Code and Error

```
PROC SQL;
  SELECT prodname, prodcost,
         manunum, manuname
    FROM PRODUCTS AS P, MANUFACTURERS AS M
      WHERE P.manunum = M.manunum;
ERROR: Ambiguous reference, column manunum is in more than one
table.
QUIT;
```

Corrective Action

To remove any and all column ambiguities, you should reference each column that appears in two or more tables with its respective table name in a SELECT statement and its clauses. For example, to reference the MANUNUM column in the MANUFACTURERS table and remove all ambiguities, you would specify the column in the SELECT statement as MANUFACTURERS.MANUNUM.

ERROR 200-322: The symbol is not recognized and will be ignored

Problem Description

In the next example, the syntax error points to the left parenthesis at the end of the second SELECT statement's WHERE clause as being invalid. The key to finding the actual problem is to work backward, line-by-line, from the point where the error is marked. Using this approach, you will notice that the second SELECT statement is a subquery (or inner query) and does not conform to valid syntax rules. As with other syntax errors, the query as well as the subquery does not execute.

Code and Error

```
PROC SQL;
  SELECT *
    FROM INVOICE
      WHERE manunum IN
        SELECT manunum
          FROM MANUFACTURERS
            WHERE UPCASE(manucity) LIKE 'SAN DIEGO%');
                                                  -
                                                  22
                                                   -
                                                   200
ERROR 22-322: Syntax error, expecting one of the following: ;,
!, !!, &, *, **, +, -, /, AND, ESCAPE, EXCEPT, GROUP, HAVING,
INTERSECT, OR, ORDER, OUTER, UNION, |, ||.

ERROR 200-322: The symbol is not recognized and will be
ignored.

QUIT;
```

Corrective Action

A subquery must conform to valid syntax rules. To correct this problem, a right parenthesis must be added at the beginning of the second SELECT statement immediately after the IN clause.

ERROR 22-322: expecting one of the following: a name, (, '.', AS, ON

Problem Description

In the next example, the syntax error identifies a WHERE clause being used in a left outer join. Although the SQL procedure permits an optional WHERE clause to be specified in the outer join syntax, an ON clause must be specified as well.

Code and Error

```
PROC SQL;
  SELECT prodname, prodtype,
         products.manunum, invenqty
    FROM PRODUCTS LEFT JOIN INVENTORY
      WHERE products.manunum =
      -----
      22
      76
ERROR 22-322: Syntax error, expecting one of the following: a
name, (, '.', AS, ON.

ERROR 76-322: Syntax error, statement will be ignored.

            inventory.manunum;
QUIT;
```

Corrective Action

To correct this problem and to conform to valid outer join syntax requirements, specify an ON clause before an optional WHERE clause that is used to subset joined results.

ERROR 180-322: Statement is not valid or it is used out of proper order

Problem Description

In the next example, the syntax error identifies the UNION statement as being invalid or used out of proper order. On further inspection, it is clear that one of two problems exists. The SQL procedure code consists of two separate queries with an invalid UNION operator specified, or a misplaced semicolon appears at the end of the first SELECT query.

Code and Error

```
PROC SQL;
  SELECT *
    FROM products
      WHERE prodcost < 300.00;
  UNION
  -----
  180
ERROR 180-322: Statement is not valid or it is used out of
proper order.
  SELECT *
    FROM products
      WHERE prodtype = 'Workstation';
QUIT;
```

Corrective Action

To correct this problem and conform to valid rules of syntax for a UNION operation, remove the semicolon after the first SELECT query.

ERROR 73-322: Expecting an AS

Problem Description

In the next example, the syntax error identifies a missing AS keyword in the CREATE VIEW statement and highlights the view's SELECT statement. If the AS keyword is not specified, the CREATE VIEW step is not executed and the view is not created.

Code and Error

```
PROC SQL;
  CREATE VIEW WORKSTATION_PRODUCTS_VIEW
    SELECT PRODNUM, PRODNAME, PRODTYPE, PRODCOST
    ------
    73
ERROR 73-322: Expecting an AS.

      FROM PRODUCTS
        WHERE UPCASE(PRODTYPE)="WORKSTATION";
QUIT;
```

Corrective Action

To correct the problem, add the AS keyword in the CREATE VIEW statement to follow valid syntax rules and define the view's query.

9.7 Summary

1. The objective of the debugging process is to identify, classify, fix, and verify a problem in a PROC SQL step, program, or application as quickly and easily as possible (see section 9.3).

2. Usage errors can cause the SAS System to stop processing, produce warnings, or produce unexpected results (see section 9.4).

3. PROC SQL provides numerous troubleshooting and debugging capabilities for the practitioner (see section 9.5).

4. PROC SQL options provide effective troubleshooting and debugging techniques for resolving coding problems (see section 9.5.2).

5. Several useful undocumented PROC SQL options are available for troubleshooting and debugging (see section 9.6).

Chapter 10

Tuning for Performance and Efficiency

10.1 Introduction

A book on PROC SQL would not be complete without some discussion of query optimization and performance. Enabling a query to run efficiently involves writing code that can take advantage of the PROC SQL query optimizer. Because PROC SQL is designed to handle a variety of processes while accommodating small to large database environments, this chapter presents a number of query tuning strategies, techniques, and options to assist you to write more efficient PROC SQL code. In this chapter you will find tips and suggestions to help identify areas where a query's inefficiencies may exist and to conduct the tuning process to achieve the best performance possible.

10.2 Understanding Performance Tuning

Performance tuning is the process of improving the way a program operates. It involves taking a program and seeing what can be done to improve performance in an intelligent, controlled manner. As you might imagine, a tuned program is one that gets the most from the existing hardware and software environment.

Performance tuning involves measuring, evaluating, and modifying a program until it uses the minimum amount of computer resources to complete its execution. The biggest problem with the tuning process is that it is sometimes difficult to determine the amount of computer resources a program uses. Complicating matters further, adequate and complete information about resource utilization is often unavailable. In fact, no simple formula exists to determine how efficiently a program runs. Often the only way to assess whether a program is running efficiently is to evaluate its performance under varying conditions, such as during interactive use or during shortages of specific resources including memory and storage.

Performance issues may be difficult to identify. It is possible to have a program that operates without any apparent problem, but does not perform as efficiently as it could. In fact a program may perform well in one environment and poorly in another. Take for example an organization that has a shortage of Direct Access Storage Device (DASD). A program that uses excessive amounts of this resource may be deemed a poor performer under these circumstances. But if the same program were run in an environment that had adequate levels of DASD, it may not be suspected or tagged as a poor performer. This distinction demonstrates the subjectivity that is frequently used to determine how a program performs and how it is linked to the specific needs (related to resource issues) an organization has at any point in time.

10.3 Sorting and Performance

Sorting data in the SQL procedure, as in other parts of the SAS System, is a CPU and memory-intensive operation. When sufficient amounts of CPU and memory resources are available, the process is usually successful. But if either of these resources is in short supply or simply not available, the sort step is doomed for failure. The first order of business for SAS users is to minimize the number of sorts in their programs. By keeping a few simple guidelines in mind, problems can be minimized.

CPU-related bottlenecks can occur if sorts are performed on disk as opposed to in memory. Because most disks are slower than physical memory, this presents an important performance issue. The most logical and efficient place to perform sorts is in memory. If the sort requires more space than can fit in available memory, the sort must be performed on disk. The objective is to determine how much space a sort will require as well as where the sort will be performed before the sort is executed.

10.3.1 User-Specified Sorting (SORTPGM= System Options)

You can control what sort utility the SAS System uses when performing sorts. By specifying the SORTPGM= system option, you can direct the SAS System to use the best possible sort utility for the environment in question. The SORTPGM= system options are displayed in the following table.

SORTPGM= System Options

Sort Option	Purpose
BEST	The BEST option uses the sort utility best suited to the data.
HOST	The HOST option tells the SAS System to use the host sort utility available on your host computer. This option may be the most efficient for large tables containing many rows of data.
SAS	The SAS option tells the SAS System to use the sort utility supplied with the SAS System.

The next example illustrates using the SORTPGM= option to select the sort utility most suited to the data. Both options use the name that is specified in the SORTNAME= option.

```
OPTIONS SORTPGM=BEST;

OPTIONS SORTPGM=HOST;
```

10.3.2 Automatic Sorting

Using the SELECT DISTINCT clause invokes an internal sort to remove duplicate rows. The single exception is when an index exists. The index is then used to eliminate the duplicate rows.

The results of a grouped query are automatically sorted using the grouping columns. When the SELECT clause contains only the columns listed in the GROUP BY clause along with any summary functions, then the duplicates in each group based on the grouping columns are removed as soon as any defined summary functions are performed. If additional columns then appear in the SELECT clause the rows are not collapsed and therefore duplicates are not removed.

10.4 Splitting Tables

Splitting tables involves moving some of the rows from one table to another table. Data is split for the purpose of separating some predetermined range of data, such as historical data from current data, so that query performance is increased. This reduces the burden imposed on queries that only access current data. The following PROC SQL example shows the current year's data being copied and then removed from a table containing five years of data.

SQL Code

```
PROC SQL;
  CREATE TABLE INVENTORY_CURRENT AS
    SELECT *
      FROM INVENTORY
        WHERE YEAR(ORDDATE) = YEAR(TODAY());

  DELETE FROM INVENTORY
    WHERE YEAR(ORDDATE) = YEAR(TODAY());
QUIT;
```

10.5 Indexes and Performance

Indexes can be used to allow rapid access to table rows. Rather than physically sorting a table (as performed by the ORDER BY clause or PROC SORT), an index is designed to set up a logical arrangement for the data without the need to physically sort it. This has the advantage of reducing CPU and memory requirements. It also reduces data access time when using WHERE clause processing (discussed in Section 10.7).

Indexes are useful, but they do have drawbacks. As data in a table is inserted, modified, or deleted, an index must be updated to address the changes. This automatic feature requires additional CPU resources to process any changes to a table. Also, as a separate structure in its own right, an index can consume considerable storage space. As a consequence, care should be exercised not to create too many indexes but to assign indexes to only those discriminating variables in a table. Here are a few suggestions for creating indexes.

- Sort data in ascending order on the key column prior to creating the index.
- Sort the data by the key variable first to achieve the greatest performance improvement.
- Sort data in ascending order by the key variable before it is appended to the table.
- Create simple indexes, when possible, to be used by most queries.
- Avoid creating one single index for all queries.
- Assign indexes to the most discriminating of variables.

- Select columns that are frequently the subject of summary functions (COUNT, SUM, AVG, MIN, MAX, etc.).
- Only create indexes that are actually needed.
- Avoid taxing CPU resources associated with index maintenance (maintaining an index during inserts, modifications, and deletions) by selecting columns that do not change frequently.
- On some operating systems indexes are stored as a separate file on disk, using additional memory and disk space to store the structure.
- To avoid excessive and unnecessary I/O operations, prior to creating an index sort data in ascending order by the most discriminating key column.
- Attempt to define composite indexes using the most discriminating of the variables as your first variable in the index.
- Select columns that do not have numerous null values because this results in a large percentage of rows with the same value.

Note: Indexes should only be created on tables where query search time needs to be optimized. Any unnecessary indexes may force the SAS System to expend resources needlessly—updating and reorganizing after insert, update, and delete operations are performed. And even worse, the SQL optimizer may accidentally use an index when it should not.

10.6 Reviewing CONTENTS Output and System Messages

While no two organizations are alike, it is not surprising to find numerous causes for a program to run at less than peak efficiency. Performance is frequently affected by the specific needs of an organization or its lack of resources. SAS users need to learn as many techniques as possible to correct problems associated with poorly performing programs. Attention should be given to individual program functions, because poor program performance often points to one or more inefficient techniques being used.

Two methods can be used to better understand potential performance issues. The first approach uses PROC CONTENTS output to examine engine/host information and library data sets (tables). The CONTENTS output provides information to determine whether a table is large enough. (The page count in the following output to show the performance improvements offered by an index). The general rule that the SQL processor adheres to is when a table is relatively small (usually fewer than three pages) there is no real advantage in using an index. In fact, using an index with a small table can actually degrade performance levels because in these situations sequential processing would be just as fast as using an index.

Results

```
                          The CONTENTS Procedure

Data Set Name       WORK.INVENTORY                      Observations          7
Member Type         DATA                                Variables             5
Engine              V9                                  Indexes               0
Created             Friday, August 20, 2004             Observation Length    20
Last Modified       Friday, August 20, 2004             Deleted Observations  0
Protection                                              Compressed            NO
Data Set Type                                           Sorted                NO
Label
Data Representation    Windows_32
Encoding               Wlatin1 Western (Windows)

                     Engine/Host Dependent Information

Data Set Page Size          4096
Number of Data Set Pages    1
First Data Page             1
Max Obs per Page            202
Obs in First Data Page      7
Number of Data Set Repairs  0
File Name                   D:\SAS Version 9.1\SAS Temporary Files
                              \_TD1632\inventory.SAS7bdat
Release Created             9.0101M0
Host Created                XP_HOME
```

(continued on next page)

-----Alphabetic List of Variables and Attributes-----

#	Variable	Type	Len	Pos	Format	Informat	Label
4	invencst	Num	6	18	DOLLAR10.2		Inventory Cost
2	invenqty	Num	3	15			Inventory Quantity
5	manunum	Num	3	24			Manufacturer Number
3	orddate	Num	4	8	MMDDYY10.	MMDDYY10.	Date Inventory Last Ordered
1	prodnum	Num	3	12			Product Number

-----Alphabetic List of Indexes and Attributes-----

#	Index	Update Centiles	Current Update Percent	# of Unique Values	Variables
1	invenqty	5	0	5	
	---				2.00036621095932
	---				-8.7692233015159E304
	---				-1.0318151782291E270
	---				5.00073295836049
	---				-8.7392210587264E304
	---				-1.4195819273297E135
	---				20.0029327871163
	---				-8.7220769199897E304

The second approach uses PROC SQL to access the dictionary tables, TABLES and COLUMNS, to determine whether a table is large enough to take advantage of the performance improvements offered by an index. See the output below.

SQL Code

```
PROC SQL;
  SELECT MEMNAME, NPAGE
    FROM DICTIONARY.TABLES
      WHERE LIBNAME='WORK' AND
        MEMNAME='INVENTORY';
```

```
   SELECT VARNUM, NAME, TYPE, LENGTH, FORMAT,
          INFORMAT, LABEL
     FROM DICTIONARY.COLUMNS
       WHERE LIBNAME='WORK' AND
         MEMNAME='INVENTORY';
QUIT;
```

Results

```
                                 The SAS System

                                                          Number
                    Member Name                         of Pages
                    ______________________________________________

                    INVENTORY                                  1

                                 The SAS System

 Column
 Number                Column   Column   Column     Column    Column
in Table   Column      Type     Length   Format     Informat  Label
______________________________________________________________________________

       1   prodnum     num           3                        Product Number
       2   invenqty    num           3                        Inventory Quantity
       3   orddate     num           4   MMDDYY10.  MMDDYY10. Date Inventory Last Ordered
       4   invencst    num           6   DOLLAR10.2           Inventory Cost
       5   manunum     num           3                        Manufacturer Number
```

The table below compares sequential table access with indexed table access. Although performance gains are data dependent, the greatest gains are realized when an index is applied to a small subset of data in a WHERE clause.

Sequential Versus Indexed Table Access

Condition	Sequential	Index
Page count < 3 pages (from CONTENTS output)	Yes	No
Small table	Yes	No
Frequent updates to table	Yes	No
Large subset of data based on WHERE processing	Yes	No
Infrequent access of table	Yes	No
Limited memory and disk space	Yes	No
Small subset of data (1% - 25% of population)	No	Yes

System messages are displayed to provide information that can help tune the indexes associated with any data sets. Setting the MSGLEVEL= system option to "I" allows the SAS System to display vital information (if available) related to the presence of one or more indexes for optimization of WHERE clause processing. With the MSGLEVEL= option turned on, the SAS log shows that the simple index INVENQTY was selected in the optimization of WHERE clause processing.

SAS Log

```
     PROC SQL;
       CREATE INDEX INVENQTY ON INVENTORY;
NOTE: Simple index invenqty has been defined.
NOTE: PROCEDURE SQL used:
      real time           0.04 seconds

       SELECT *
         FROM INVENTORY
           WHERE invenqty < 3;
INFO: Index invenqty selected for WHERE clause
optimization.
     QUIT;
NOTE: PROCEDURE SQL used:
      real time           0.65 seconds
```

10.7 Optimizing WHERE Clause Processing with Indexes

To get the best possible performance from programs containing SQL procedure code, an index and WHERE clause can be used together (see the list below). Using a WHERE clause restricts processing in a table to a subset of selected rows (see Chapter 2, "Working with Data in PROC SQL" for specific details). When an index exists, the SQL processor determines whether to take advantage of it during WHERE clause processing. Although the SQL processor determines whether using an index will ultimately benefit performance, when it does the result can be an improvement in processing speeds.

- Comparison operators such as EQ (=), LT (<), GT (>), LE (<=), GE (>=), and NOT
- Comparison operators with the IN operator
- Comparison operators with the colon modifier (for example, NOT = :"Ab")
- CONTAINS operator
- IS NULL or IS MISSING operator
- Pattern-matching operators such as LIKE and NOT LIKE

10.7.1 Constructing Efficient Logic Conditions

When constructing a chain of AND conditions in a WHERE clause, specify the most restrictive conditional values first. This way the SQL processor expends fewer resources by bypassing rows that do not satisfy the first conditional value in the WHERE clause. For example, the first PROC SQL step below may expend more resources because the first condition "SOFTWARE" occurs even for products costing more than $99.00.

SQL Code (Less Efficient)

```
PROC SQL;
  SELECT *
    FROM PRODUCTS
      WHERE UPCASE(PRODTYPE) = 'SOFTWARE' AND
            PRODCOST < 100.00;
QUIT;
```

For this data, a more efficient way of producing the same results as the previous example, while reducing CPU resources, is to code the second and more restrictive condition first so it appears as follows.

SQL Code (More Efficient)

```
PROC SQL;
  SELECT *
    FROM PRODUCTS
      WHERE PRODCOST < 100.00 AND
            UPCASE(PRODTYPE) = 'SOFTWARE';
QUIT;
```

Another popular construct uses a series of OR conditions equality tests or the IN predicate to select rows that match the multiple conditions. Programmers often order these kinds of lists by order of magnitude or alphabetically to make the lists easier to maintain. A better and more efficient way would be to order the list from the most frequently occurring values to the least frequent.

Note: One way to determine the frequency of values is to submit the following code: `SELECT COUNT(PRODTYPE) FROM SQL.PRODUCTS GROUP BY PRODTYPE;`

Once the frequencies are known, they can be specified in that order. This way the SQL processor expends fewer resources locating frequently occurring values because it has to perform fewer steps to return a value of TRUE. For example, the first SQL step below expends more resources because the first condition "LAPTOP" occurs less frequently than the value "SOFTWARE". Consequently, the SQL processor needs to process the second condition in order to find a match resulting in a value of TRUE being returned.

SQL Code (Less Efficient):

```
PROC SQL;
  SELECT *
    FROM PRODUCTS
      WHERE UPCASE(PRODTYPE) IN ('LAPTOP', 'SOFTWARE');
QUIT;
```

A more efficient way of processing the same data but generating the same results as the previous code is to place “SOFTWARE” first as follows:

SQL Code (More Efficient):

```
PROC SQL;
  SELECT *
    FROM PRODUCTS
      WHERE UPCASE(PRODTYPE) IN ('SOFTWARE', 'LAPTOP');
QUIT;
```

10.7.2 Avoiding UNIONs

UNIONs are executed by creating two internal sets, then merge-sorting the results together. Duplicate rows are automatically eliminated from the final results. For example, the SQL procedure code illustrated below first constructs the two result sets from each query, then merges and sorts the two sets together, and eliminates duplicate rows from the final results.

SQL Code (Less Efficient):

```
PROC SQL;
  SELECT *
    FROM PRODUCTS
      WHERE UPCASE(PRODTYPE) = 'LAPTOP'

  UNION

  SELECT *
    FROM PRODUCTS
      WHERE UPCASE(PRODTYPE) = 'SOFTWARE';
QUIT;
```

To improve UNION performance, SQL procedure code can be converted to a single query using OR conditions in a WHERE clause. The next example illustrates the previous SQL procedure code being made more efficient by converting the UNION to a single query using an **OR** operator (or **IN** predicate) in a WHERE clause.

SQL Code (More Efficient):

```
PROC SQL;
  SELECT DISTINCT *
    FROM PRODUCTS
      WHERE UPCASE(PRODTYPE) = 'SOFTWARE' OR
            UPCASE(PRODTYPE) = 'LAPTOP';
QUIT;
```

<or>

```
PROC SQL;
  SELECT DISTINCT *
    FROM PRODUCTS
      WHERE UPCASE(PRODTYPE) IN ('SOFTWARE', 'LAPTOP');
QUIT;
```

Another approach that can aid in improving the way a UNION performs is to specify the ALL keyword with the UNION operator, as long as duplicates are not an issue. Because UNION ALL does not remove duplicate rows, CPU resources may be improved. The next example shows the **UNION ALL** coding construct being used to perform what amounts to an append operation, thereby bypassing the sort altogether because the duplicate rows are not removed.

SQL Code (Less Efficient):

```
PROC SQL;
  SELECT *
    FROM PRODUCTS
      WHERE UPCASE(PRODTYPE) = 'LAPTOP'

  UNION ALL

  SELECT *
    FROM PRODUCTS
      WHERE UPCASE(PRODTYPE) = 'SOFTWARE';
QUIT;
```

10.8 Summary

1. Performance tuning involves measuring, evaluating, and modifying a query's execution to achieve an optimal balance between competing computer resources (see section 10.2).

2. Avoid specifying an ORDER BY clause when creating a table or view (see section 10.3).

3. When sorting is necessary, specify the SORTPGM= system option to instruct the SAS System to use the best possible sort utility relative to the size of the database environment (see section 10.3.1).

4. Care should be exercised to assign indexes to only those discriminating variables in a table and to avoid creating too many indexes (see section 10.5).

5. There is no advantage in creating or using an index when a table is relatively small (usually fewer than three pages) (see section 10.6).

6. Setting the MSGLEVEL= system option to "I" allows the SAS System to display vital information (if available) relative to the presence of one or more indexes for optimization of WHERE clause processing (see section 10.6).

7. Apply WHERE clause processing to restrict the number of rows of the result table (see section 10.7).

8. When constructing a chain of AND conditions in a WHERE clause, specify the most restrictive conditional values first (see section 10.7.1).

Glossary

Term	Definition
alias	a temporary, alternate name assigned to a table for use in a join query.
base table	a primary table.
CALCULATED column	a column that is used in the same SELECT statement in which the calculation occurs.
Cartesian product	a type of join that yields all the possible combinations of rows and columns.
case expression	an expression that specifies relations among columns and values in conditional logic. Similar to IF-THEN logic.
clause	part of an SQL statement, beginning with a keyword such as SELECT and generally containing an argument.
column	a field or variable in a table. Each column has a unique name, contains data of a specific type, and has certain attributes.
column-definition	refers to a column name and column attributes including type, length, informat, format, and label information.

Term	Definition
composite index	a separate SAS file (stored in the same SAS library the data resides in) consisting of directions that permit access to rows of data based on two or more columns within a table.
concatenation operator	an operator that allows two or more character strings to be concatenated together forming a single string.
correlated subquery	a subquery that evaluates multiple times, once for each row of data accessed by the main query.
Data Definition Language (DDL)	statements that define (redefine) the definition of one or more existing tables.
database design	the process of identifying end-user requirements and defining the structure of data values on a physical level.
DISTINCT keyword	a keyword that selects duplicate values in a column only one time and causes the SQL procedure to remove duplicate rows from the output.
equijoin	a join in which all the columns in two or more tables being joined are included in the results based on the comparison of columns being equal.
index	a data structure that permits a row of data in a table to be located without having to read the entire table.
intersect	an output table consisting of all the unique rows from the intersection of two query expressions.
join	the combination of data from two or more tables to produce a single result table.
key	identification of a row of data by the columns in an index.
missing value	a term that describes the contents of a variable that contains no data for a particular row or observation. A missing value is equivalent to an SQL null value.
nesting	views that are dependent on other views.
normalization	the process of ensuring that a database does not contain redundant information in two or more of its tables.

Term	Definition
null value	a special value that indicates the absence of information. Analogous to missing values.
ODS	the Output Delivery System provides formatting engines to automatically convert monospace output into more readable, proportionally spaced output.
operator	any of several symbols that request a comparison, a logical operation, or an arithmetic calculation.
outer join	an asymmetric join where rows are selected unproportionally from its parts. There are three kinds: left, right, and full. See also join.
pattern matching	technique used to find patterns in data.
performance tuning	process of improving the efficiency of a query, join, insert, update, or delete operation in a way to reduce the unnecessary expenditure of computer resources.
phonetic matching	the sounds-like operator searches and selects character data that sound alike or have spelling variations.
predicate	an object used to perform direct comparisons between two conditions or expressions.
query	a set of instructions that requests particular information from one or more data sources.
row	the horizontal component of a table.
scalar expression	a quantity represented by a single number or value.
self-join	a single table that is joined with itself.
set operator	an operator that combines or merges separate queries together.
simple index	a separate SAS file (stored in the same SAS library the data resides in) consisting of directions that permit access to rows of data based on a single column within a table.

Term	Definition
sounds-like operator	see phonetic matching.
subquery	a query that is embedded in a WHERE clause of the main query.
symmetric join	an equal selection from two parts.
union	an output table consisting of all the unique rows from the combination of query expressions.
updatable view	a view that is able to update a physical table.
view	a virtual table (non-existent) that contains a set of instructions on how to obtain data from base tables.
virtual table	a virtual table does not contain any data but is a query consisting of a set of instructions.

References

Bowman, Judith S., Sandra L. Emerson, and Marcy Darnovsky. 1993. *The Practical SQL Handbook: Using Structured Query Language* (2d Edition). Reading, MA: Addison-Wesley Publishing Company, Inc.

Celko, Joe. 1995. *Instant SQL Programming*. Birmingham, UK: Wrox Press Ltd.

Celko, Joe. 1995. *SQL for Smarties: Advanced SQL Programming*. San Francisco, CA: Morgan Kaufmann Publishers, Inc.

Celko, Joe. 1999. *SQL for Smarties: Advanced SQL Programming* (2d Edition). San Francisco, CA: Morgan Kaufmann Publishers, Inc.

Church, Lewis, Jr. 1999. "Performance Enhancements to PROC SQL in Version 7 of the SAS System." *Proceedings of the Twenty-Fourth Annual SAS Users Group International Conference,* Miami Beach, Florida, 336-341.

Date, C.J., and Hugh Darwen. 1993. *A Guide to the SQL Standard* (3d Edition). Reading, MA: Addison-Wesley Publishing Company, Inc.

Dorfman, Paul M. 2001. "Table Look-up by Direct Addressing: Key-Indexing—Bitmapping—Hashing," *Proceedings of the Twenty-Sixth Annual SAS Users Group International Conference*, Long Beach, CA, Paper 8-26.

Gruber, Martin. 1990. *Understanding SQL*. Alameda, CA: Sybex, Inc.

Lafler, Kirk Paul. 1994. "Joining Tables of Data in the SQL Procedure," *Proceedings of the Second Annual Western Users of SAS Software Regional Users Group Conference* 303-308.

Lafler, Kirk Paul. 1995. "Frame Your View of Data with the SQL Procedure." *Proceedings of the Third Annual Southeast SAS Users Group Conference,* Savannah, GA, 34-37.

Lafler, Kirk Paul. 1998. "Ten Great Reasons to Learn SAS Software's SQL Procedure." *Proceedings of the Twenty-Third Annual SAS Users Group International Conference*, Nashville, TN, 738-743.

Lafler, Kirk Paul. 2002. "A Visual Introduction to SQL Joins." *Proceedings of the Twenty-Seventh Annual SAS Users Group International Conference*, Orlando, FL, Paper 72-27.

Lafler, Kirk Paul. 2003a. "Undocumented and Hard-to-Find PROC SQL Features." *Proceedings of the Eleventh Annual Western Users of SAS Software Conference*, San Francisco, CA, Paper 19-28.

Lafler, Kirk Paul. 2003b. "Intermediate and Advanced PROC SQL." *Proceedings of the Twenty-Eighth SAS Users Group International Conference*, Seattle, WA, Sunday Seminar.

Lafler, Kirk Paul. 2004. "Efficiency Techniques for Beginning PROC SQL Users." *Proceedings of the Twenty-Ninth SAS Users Group International Conference*, San Francisco, CA, Paper 127-29.

Lafler, Kirk Paul, and Bruce Jeffrey Lafler. 1995. "Diving into the SAS System with the SQL Procedure." *Proceedings of the Twentieth Annual SAS Users Group International Conference*, Orlando, FL, 1076-1081.

Lorie, Raymond A., and Jean-Jacques Daudenarde. 1991. *SQL & Its Applications*. Englewood Cliffs. NJ: Prentice Hall, Inc.

Melton, Jim, and Alan R. Simon. 1992. *Understanding the New SQL—A Complete Guide*. San Francisco, CA: Morgan Kaufmann Publishers, Inc.

Morgan, Bryan, and Jeff Perkins. 1995. *Teach Yourself SQL in 14 Days*. Indianapolis, IN: Sams Publishing.

Olson, Diane. 2000. "Power Indexing: A Guide to Using Indexes Effectively in Nashville Releases." *Proceedings of the Eighth Annual Western Users of SAS Software Regional Users Group Conference*, Scottsdale, AZ, 207-211.

SAS Institute Inc. 1989. *SAS Guide to the SQL Procedure: Usage and Reference, Version 6*, First Edition. Cary, NC: SAS Institute Inc.

SAS Institute Inc. 2000. SAS Technical Support Document TS-320: *Inside PROC SQL's Query Optimizer*. Cary, NC: SAS Institute Inc.

SAS Institute Inc. 2000. SAS Technical Support Document TS-553: *SQL Joins—The Long and the Short of It*. Cary, NC: SAS Institute Inc.

SAS Institute Inc. 2000. *SAS SQL Procedure User's Guide, Version 8.* Cary, NC: SAS Institute Inc.

Trimble, J. Harvey, and David Chappell. 1989. *A Visual Introduction to SQL*. New York, NY: John Wiley & Sons, Inc.

Index

A

E

N

T

U

V

W

X

Y

Z

Special Characters

Call your local SAS office to order these books from Books by Users Press

Advanced Log-Linear Models Using SAS®
by **Daniel Zelterman** . . . . Order No. A57496

Analysis of Clinical Trials Using SAS®: A Practical Guide
by **Alex Dmitrienko, Walter Offen, Christy Chuang-Stein,**
and **Geert Molenbergs** . . . . Order No. A59390

Annotate: Simply the Basics
by **Art Carpenter** . . . . Order No. A57320

Applied Multivariate Statistics with SAS® Software, Second Edition
by **Ravindra Khattree**
and **Dayanand N. Naik** . . . . Order No. A56903

Applied Statistics and the SAS® Programming Language, Fourth Edition
by **Ronald P. Cody**
and **Jeffrey K. Smith** . . . . Order No. A55984

An Array of Challenges — Test Your SAS® Skills
by **Robert Virgile** . . . . Order No. A55625

Carpenter's Complete Guide to the SAS® Macro Language, Second Edition
by **Art Carpenter** . . . . Order No. A59224

The Cartoon Guide to Statistics
by **Larry Gonick**
and **Woollcott Smith** . . . . Order No. A5515

Categorical Data Analysis Using the SAS® System, Second Edition
by **Maura E. Stokes, Charles S. Davis,**
and **Gary G. Koch** . . . . Order No. A57998

Cody's Data Cleaning Techniques Using SAS® Software
by **Ron Cody** . . . . Order No. A57198

Common Statistical Methods for Clinical Research with SAS® Examples, Second Edition
by **Glenn A. Walker** . . . . Order No. A58086

Debugging SAS® Programs: A Handbook of Tools and Techniques
by **Michele M. Burlew** . . . . Order No. A57743

Efficiency: Improving the Performance of Your SAS® Applications
by **Robert Virgile** . . . . Order No. A55960

Genetic Analysis of Complex Traits Using SAS®
Edited by **Arnold M. Saxton** . . . . Order No. A59454

A Handbook of Statistical Analyses Using SAS®, Second Edition
by **B.S. Everitt**
and **G. Der** . . . . Order No. A58679

Health Care Data and the SAS® System
by **Marge Scerbo, Craig Dickstein,**
and **Alan Wilson** . . . . Order No. A57638

The How-To Book for SAS/GRAPH® Software
by **Thomas Miron** . . . . Order No. A55203

Instant ODS: Style Templates for the Output Delivery System
by **Bernadette Johnson** . . . . Order No. A58824

In the Know... SAS® Tips and Techniques From Around the Globe
by **Phil Mason** . . . . Order No. A55513

support.sas.com/pubs

Integrating Results through Meta-Analytic Review Using SAS® Software
by **Morgan C. Wang**
and **Brad J. Bushman** . . . . Order No. A55810

Learning SAS® in the Computer Lab, Second Edition
by **Rebecca J. Elliott**. . . . . Order No. A57739

The Little SAS® Book: A Primer
by **Lora D. Delwiche**
and **Susan J. Slaughter** . . . . Order No. A55200

The Little SAS® Book: A Primer, Second Edition
by **Lora D. Delwiche**
and **Susan J. Slaughter** . . . . Order No. A56649
(updated to include Version 7 features)

The Little SAS® Book: A Primer, Third Edition
by **Lora D. Delwiche**
and **Susan J. Slaughter** . . . . Order No. A59216
(updated to include SAS 9.1 features)

Logistic Regression Using the SAS® System: Theory and Application
by **Paul D. Allison** . . . . Order No. A55770

Longitudinal Data and SAS®: A Programmer's Guide
by **Ron Cody** . . . . Order No. A58176

Maps Made Easy Using SAS®
by **Mike Zdeb**. . . . . Order No. A57495

Models for Discrete Data
by **Daniel Zelterman** . . . . Order No. A57521

Multiple Comparisons and Multiple Tests Using SAS® Text and Workbook Set
(*books in this set also sold separately*)
by **Peter H. Westfall, Randall D. Tobias, Dror Rom, Russell D. Wolfinger,**
and **Yosef Hochberg** . . . . Order No. A55770

Multiple-Plot Displays: Simplified with Macros
by **Perry Watts** . . . . Order No. A58314

Multivariate Data Reduction and Discrimination with SAS® Software
by **Ravindra Khattree,**
and **Dayanand N. Naik** . . . . Order No. A56902

support.sas.com/pubs

Output Delivery System: The Basics
by **Lauren E. Haworth** . . . . Order No. A58087

Painless Windows: A Handbook for SAS® Users, Third Edition
by **Jodie Gilmore** . . . . Order No. A58783
(*updated to include Version 8 and SAS 9.1 features*)

PROC TABULATE by Example
by **Lauren E. Haworth** . . . . Order No. A56514

Professional SAS® Programming Shortcuts
by **Rick Aster** . . . . Order No. A59353

Quick Results with SAS/GRAPH® Software
by **Arthur L. Carpenter**
and **Charles E. Shipp**. . . . . Order No. A55127

Quick Results with the Output Delivery System
by **Sunil Gupta** . . . . Order No. A58458

Quick Start to Data Analysis with SAS®
by **Frank C. Dilorio**
and **Kenneth A. Hardy** . . . . Order No. A55550

Reading External Data Files Using SAS®: Examples Handbook
by **Michele M. Burlew** . . . . Order No. A58369

Regression and ANOVA: An Integrated Approach Using SAS® Software
by **Keith E. Muller**
and **Bethel A. Fetterman** . . . . Order No. A57559

SAS® Applications Programming: A Gentle Introduction
by **Frank C. Dilorio** . . . . Order No. A56193

SAS® for Forecasting Time Series, Second Edition
by **John C. Brocklebank**
and **David A. Dickey** . . . . Order No. A57275

SAS® for Linear Models, Fourth Edition
by **Ramon C. Littell, Walter W. Stroup,**
and **Rudolf Freund** . . . . Order No. A56655

SAS® for Monte Carlo Studies: A Guide for Quantitative Researchers
by **Xitao Fan, Ákos Felsővályi, Stephen A. Sivo,**
and **Sean C. Keenan**. . . . . Order No. A57323

SAS® Functions by Example
by **Ron Cody** . . . . Order No. A59343

SAS® Macro Programming Made Easy
by **Michele M. Burlew** . . . . Order No. A56516

SAS® Programming by Example
by **Ron Cody**
and **Ray Pass** . . . . Order No. A55126

SAS® Programming for Researchers and Social Scientists, Second Edition
by **Paul E. Spector** . . . . Order No. A58784

SAS® Survival Analysis Techniques for Medical Research, Second Edition
by **Alan B. Cantor** . . . . Order No. A58416

SAS® System for Elementary Statistical Analysis, Second Edition
by **Sandra D. Schlotzhauer**
and **Ramon C. Littell**. . . . . Order No. A55172

SAS® System for Mixed Models
by **Ramon C. Littell, George A. Milliken, Walter W. Stroup,** and **Russell D. Wolfinger** . . Order No. A55235

SAS® System for Regression, Second Edition
by **Rudolf J. Freund**
and **Ramon C. Littell**. . . . . Order No. A56141

SAS® System for Statistical Graphics, First Edition
by **Michael Friendly** . . . . Order No. A56143

The SAS® Workbook and Solutions Set
(*books in this set also sold separately*)
by **Ron Cody** . . . . Order No. A55594

Selecting Statistical Techniques for Social Science Data: A Guide for SAS® Users
by **Frank M. Andrews, Laura Klem, Patrick M. O'Malley, Willard L. Rodgers, Kathleen B. Welch,**
and **Terrence N. Davidson** . . . . Order No. A55854

Statistical Quality Control Using the SAS® System
by **Dennis W. King**. . . . . Order No. A55232

A Step-by-Step Approach to Using the SAS® System for Factor Analysis and Structural Equation Modeling
by **Larry Hatcher**. . . . . Order No. A55129

A Step-by-Step Approach to Using the SAS® System for Univariate and Multivariate Statistics
by **Larry Hatcher**
and **Edward Stepanski** . . . . Order No. A55072

Step-by-Step Basic Statistics Using SAS®: Student Guide and Exercises
(*books in this set also sold separately*)
by **Larry Hatcher**. . . . . Order No. A57541

Survival Analysis Using the SAS® System: A Practical Guide
by **Paul D. Allison** . . . . Order No. A55233

Tuning SAS® Applications in the OS/390 and z/OS Environments, Second Edition
by **Michael A. Raithel** . . . . Order No. A58172

Univariate and Multivariate General Linear Models: Theory and Applications Using SAS® Software
by **Neil H. Timm**
and **Tammy A. Mieczkowski** . . . . Order No. A55809

Using SAS® in Financial Research
by **Ekkehart Boehmer, John Paul Broussard,**
and **Juha-Pekka Kallunki** . . . . Order No. A57601

Using the SAS® Windowing Environment: A Quick Tutorial
by **Larry Hatcher**. . . . . Order No. A57201

Visualizing Categorical Data
by **Michael Friendly** . . . . Order No. A56571

Web Development with SAS® by Example
by **Frederick Pratter** . . . . Order No. A58694

Your Guide to Survey Research Using the SAS® System
by **Archer Gravely**. . . . . Order No. A55688

support.sas.com/pubs

JMP® Books

JMP® Start Statistics, Third Edition
by **John Sall, Ann Lehman,**
and **Lee Creighton** . . . . Order No. A58166

Regression Using JMP®
by **Rudolf J. Freund, Ramon C. Littell,**
and **Lee Creighton** . . . . Order No. A58789

support.sas.com/pubs